Structure–Property Relationships in Polymers

Structure–Property Relationships in Polymers

Raymond B. Seymour

College of Science and Technology
University of Southern Mississippi
Hattiesburg, Mississippi

and

Charles E. Carraher, Jr.

College of Science and Engineering
Wright State University
Dayton, Ohio

Plenum Press • New York and London

Library of Congress Cataloging in Publication Data

Seymour, Raymond Benedict, 1912–
 Structure–property relationships in polymers.

 Bibliography: p.
 Includes index.
 1. Polymers and polymerization. I. Carraher, Charles E. II. Title.
TA455.P58S48   1984   620.1′92   84-13475
ISBN 0-306-41650-6

© 1984 Plenum Press, New York
A Division of Plenum Publishing Corporation
233 Spring Street, New York, N.Y. 10013

Printed in the United States of America

* | Preface

The first concern of scientists who are interested in synthetic polymers has always been, and still is: How are they synthesized? But right after this comes the question: What have I made, and for what is it good? This leads to the important topic of the structure–property relations to which this book is devoted.

Polymers are very large and very complicated systems; their characterization has to begin with the chemical composition, configuration, and conformation of the individual molecule. The first chapter is devoted to this broad objective. The immediate physical consequences, discussed in the second chapter, form the basis for the physical nature of polymers: the supermolecular interactions and arrangements of the individual macromolecules.

The third chapter deals with the important question: How are these chemical and physical structures experimentally determined? The existing methods for polymer characterization are enumerated and discussed in this chapter.

The following chapters go into more detail.

For most applications—textiles, films, molded or extruded objects of all kinds—the mechanical and the thermal behaviors of polymers are of preponderant importance, followed by optical and electric properties. Chapters 4 through 9 describe how such properties are rooted in and dependent on the chemical structure.

More-detailed considerations are given to certain particularly important and critical properties such as the solubility and permeability of polymeric systems.

Macromolecules are not always the final goal of the chemist—they may act as intermediates, reactants, or catalysts. This topic is presented in Chapters 10 and 11.

The last five chapters go somewhat deeper into particularly interesting and important groups of macromolecules such as polyolefins, vinyls, styrenics, and aromatic polymers.

Special chapters are devoted to the important role which certain additives—stabilizers, plasticizers, flame retardants, and others—play in the ultimate behavior of polymeric systems and to the class of high-performance polymers, the study of which is a new but rapidly growing branch of polymer science, which will play a growing role in the construction of space vehicles, airplanes, ships, and all vehicles of ground transportation.

H. Mark

* | Acknowledgments

The authors are extremely appreciative of the advice provided by Drs. Herman F. Mark, Rudolph Deanin, and Roger Porter who reviewed the manuscript. These scientists are recognized internationally for their contributions to polymer science education at the Polytechnic Institute of New York and the Universities of Lowell and Massachusetts respectively. Appreciation is also given to Charles E. Carraher III and Shawn Carraher for their assistance in the proofing and indexing of the text.

R.B.S.
C.E.C.

✱ | Contents

13.   Polymeric Hydrocarbons with Pendant Groups

14.   Aliphatic Polymers with Heteroatom Chains

15. High-Performance Polymers

16. Selection of Polymers for Special Applications

1 | Chemical Structure of Polymers

1.1 Introduction

Polymers such as proteins and nucleic acids were present when the first unicellular organism appeared on earth. Other natural polymers, such as cellulose and starch, have been utilized for food, shelter, and clothing for thousands of years. Cellulose, polyisoprene, and shellac were converted to useful man-made plastics, fibers, and elastomers in the 19th century, but these conversions were based primarily on empirical knowledge.

Nobel laureate Emil Fischer elucidated the chemical structure of carbohydrates and proteins, and Leo Baekeland produced the first commercial synthetic polymer in the early part of the 20th century. However, these and many other synthetic polymers developed in the first part of the 20th century were produced without much knowledge of polymer structure.

Although the essential concept of macromolecular structure that is accepted today was proposed by Nobel laureate Hermann Staudinger in the 1920s, few organic chemists then accepted it. Today the concept that linear polymers consist of very long chains of covalently bonded atoms is almost universally accepted.[1] It is now recognized that regardless of whether a polymer is natural or synthetic, a fiber, a plastic, an elastomer, a coating, or an adhesive, it consists of a backbone of multitudinous covalently bonded atoms. The differences between fibers, plastics, and elastomers depend, to a large extent, on the regularity of polymer structure and the relative strength of the intermolecular secondary valence bonds. The latter may be strong hydrogen bonds, moderately strong dipole–dipole forces, or London forces.

$$
\begin{array}{cc}
\text{H} & \text{H} \\
| & | \\
\text{C} - \text{C} \\
| & | \\
\text{H} & \text{H}
\end{array}\Big)_n
$$

The repeating ethylene unit shown above may be used to illustrate the simplest (structure-wise) macromolecule. Linear polyethylene (HDPE) consists of many such ethylene units joined covalently by strong primary valence bonds in a continuous linear chain. The repeating unit is called a *monomer* after the Greek *monos*, meaning "single," and the long chain of repeating units is called a *polymer* after the Greek *poly*, meaning "many." Some purists prefer the term *macromolecule*, meaning "large molecule," rather than polymer. However, the two terms, macromolecule and polymer, are used interchangeably in this book.

The length of the polymer chain in HDPE and other polymers is dependent on the number of repeating units, or *mers*, present in the polymer chain; this number is designated by the letter n or the abbreviation DP (degree of polymerization). The molecular weight, or molar mass, of the polymer, M, is of course equal to n times the molecular weight of the repeating unit, m, i.e., $M = nm$. Thus if n for HDPE $(CH_2CH_2)_n$ is equal to 1000, M will be equal to 2800. Because the intermolecular forces are weak London forces, the attractions between the polymer chains are relatively weak. The term *oligomer*, derived from the Greek *oligo*, meaning "few," is used to describe relatively low molecular weight polymers having n values less than 20.

Compounds with molecular structures similar to those of the repeating units are called model compounds, and much information about the properties of polymers may be derived from knowledge of these model compounds. Thus, a person who knows the chemical properties of ethane can fairly well extrapolate this to the chemistry of HDPE. Of course, the physical and thermal properties of HDPE are much different from those of ethane.

1.2 Shapes and Energy Considerations

The bond angles of the covalently bonded tetrahedral carbon atoms in both the model hydrocarbon compounds and the corresponding polymers are 109°28′, and the lengths of the C—H and C—C bonds in HDPE are 0.109 and 0.154 nm, respectively. The C—H and C—C bond energies are about 98 and 80 kcal/mol, respectively.

Both the C—H and the C—C bonds contribute to the size of the HDPE molecule, but the total length of about 1000 C—C bonds is much more significant than the length of a single C—H bond. Hence the dimensions of an HDPE polymer with a DP of 1000 may be likened to those of a rope with a length which is over 500 times its diameter (Figure 1.1).

The carbon–carbon (C—C) bonds in HDPE are flexible because of the so-called free rotation of these bonds. Because of this so-called free rotation, the polymer chains are in a constant state of micro-Brownian motion or a wiggling segmental motion at temperatures above the glass transition temperature T_g, which is defined as the temperature at which this segmental motion begins as the temperature of the polymer is raised. It is important to note that this segmental motion occurs without any significant change in the characteristic bond angle of 109°28'. The speed of these intramolecular motions, which result in a myriad of different shapes or conformations, is rapid and temperature dependent.

It should also be noted that there is a slight preference for chain conformations of lower energy. Newman projections for the two extreme conformations of ethane are shown in Figure 1.2. The value of 2.8 kcal/mol represents the energy opposing free rotation about the C—C bond.

The intramolecular mobility, which yields many other conformations between these two extremes, is the basis for viscoelasticity in polymers. Pendant groups, such as the methyl group in isobutane, and lower temperatures decrease the mobility of model compounds such as ethane as well as the polymer chains of polyethylene.

The crystallinity of a polymer such as polyethylene typically increases as the molecular weight and the structural regularity increase but decreases as the extent of irregular branching in the polymer molecule increases. Thus because of its regular structure, HDPE, like linear paraffins, readily forms crystals. In contrast, branched or low-density polyethylene (LDPE) is less crystalline because of its more irregular structure.

LDPE, the original commercial polyethylene, is a highly branched polymer

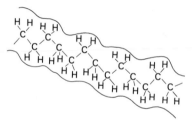

Figure 1.1. A portion of HDPE illustrating the ropelike characteristics.

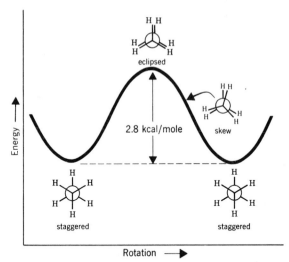

Figure 1.2. Potential energy profile illustrating the potential energy changes associated with rotation about the C—C bond of ethane.

with a relatively higher volume per unit and a lower specific gravity (0.92) than HDPE. In contrast, commercial HDPE, produced originally by Nobel laureate Karl Ziegler, has a specific gravity of 0.96. Linear low-density polyethylene (LLDPE) can be considered as a copolymer of ethylene and higher-molecular-weight linear alkenes such as 1-butene.

Simulated representations of linear (HDPE) and branched (LDPE) polyethylene are given in Figures 1.3 and 1.4.

As shown originally by Malcolm Dole, polyethylene molecules may be cross-linked when subjected to high-energy radiation. These three-dimensional network polymers may be represented by the structure shown in Figure 1.5.

Two terms, *configuration* and *conformation,* are often confused. *Configuration* refers to arrangements fixed by chemical bonding, which cannot be

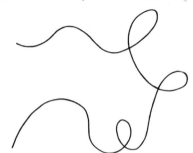

Figure 1.3. Linear polyethylene (HDPE).

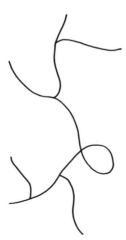

Figure 1.4.   Branched polyethylene (LDPE).

altered except through primary bond breakage. Terms such as *head-to-tail, d* and *l, cis* and *trans* refer to the configuration of a chemical species. *Conformation* refers to arrangements about single primary bonds. Polymers in solution or in melts continuously undergo conformational changes.

Nobel laureate Giulio Natta produced commercial rigid polymers of propylene in 1957 by the use of a coordination catalyst system similar to that used previously by Karl Ziegler for the production of HDPE. Commercial polypropylene (PP) is a linear polymer with regularly spaced methyl (CH_3, Me) pendant groups on every other carbon atom in the polymer chain. This alternation of carbons containing a methyl group as shown in Figure 1.6 is called a head-to-tail configuration, and it is typical for polymers. The nontypical configurations—head-to-head, tail-to-tail—is shown in Figure 1.7. Structures of other common vinyl polymers appear in Figure 1.8.

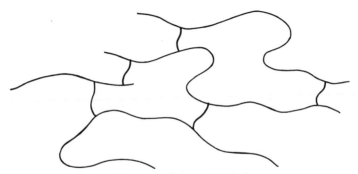

Figure 1.5.   Cross-linked polyethylene.

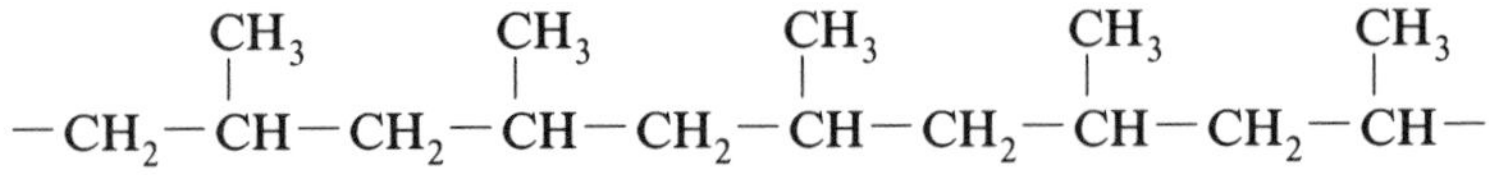

$$
\begin{array}{ccccc}
\mathrm{CH_3} & \mathrm{CH_3} & \mathrm{CH_3} & \mathrm{CH_3} & \mathrm{CH_3} \\
| & | & | & | & | \\
-\mathrm{CH_2}-\mathrm{CH}-\mathrm{CH_2}-\mathrm{CH}-\mathrm{CH_2}-\mathrm{CH}-\mathrm{CH_2}-\mathrm{CH}-\mathrm{CH_2}-\mathrm{CH}-
\end{array}
$$

Figure 1.6. Portion of PP illustrating the head-to-tail configuration, which is typical for polymers.

Configuration also refers to structural regularity with respect to the substituted carbon within the polymer chain (Figure 1.9). For linear homopolymers derived from monomers of the form $H_2C = CHX$, such as styrene and vinyl chloride, configuration from monomeric unit to monomeric unit can vary somewhat randomly (*atactic*; abbreviated *at*) with respect to the geometry and configuration about the carbon to which the X is attached, or can vary alternately (*syndiotactic*; *st*), or can be uniform (*isotactic*; *it*) when all the X groups can be placed on the same side of a backbone plane.

The term *tactic* is taken from the Greek word *taktikos*, meaning "placed in order." The prefixes *iso, syndio,* and *a* mean "same," "two together," and "without," respectively.

Enantiomers, nonsuperimposable mirror-image isomers, could also be designated as *dd* or *ll* instead of *it*; *dldl* instead of *st*; and, for example, *dllddl-dlldld* instead of *at*. The isotacticity of commercially produced PP may be increased by the hexane extraction of *at*-PP, but most modern PP meets specifications for a high degree of isotacticity without the solvent extraction step.

Because of its irregular structure, *at*-PP is an amorphous polymer with a softening point lower than that of *it*-PP. In contrast, because of its regular structure, commercial *it*-PP is a higher-melting crystalline solid. It is important to note that similar stereochemical concepts apply to other vinyl polymers with pendant groups, such as polyvinyl chloride (PVC) and polystyrene (PS).

Another type of stereogeometry arises in the case of 1,4 dienes such as 1,4-butadiene, in which rotation in the polymer is restricted because of the presence of the double bond. Polymerization can occur through a single, static double bond, producing 1,2 products which can exist in the stereoregular isotactic and syndiotactic forms, and in the irregular atactic form. The stereoregular forms are rigid, crystalline materials, whereas the atactic forms are soft elastomers.

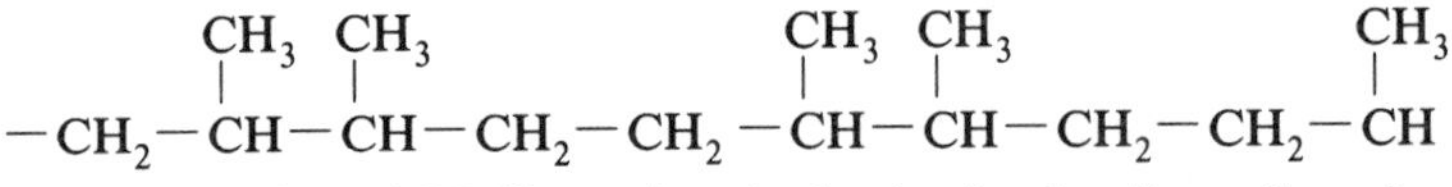

$$
\begin{array}{cccccc}
\mathrm{CH_3} & \mathrm{CH_3} & & \mathrm{CH_3} & \mathrm{CH_3} & & \mathrm{CH_3} \\
| & | & & | & | & & | \\
-\mathrm{CH_2}-\mathrm{CH}-\mathrm{CH}-\mathrm{CH_2}-\mathrm{CH_2}-\mathrm{CH}-\mathrm{CH}-\mathrm{CH_2}-\mathrm{CH_2}-\mathrm{CH}
\end{array}
$$

Figure 1.7. Portion of PP illustrating the head-to-head, tail-to-tail configuration, which is nontypical for polymers.

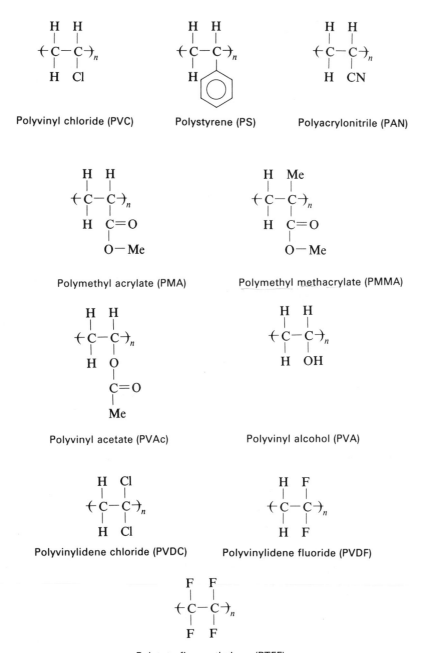

Figure 1.8. Structures of the basic repeating units for select common vinyl polymers.

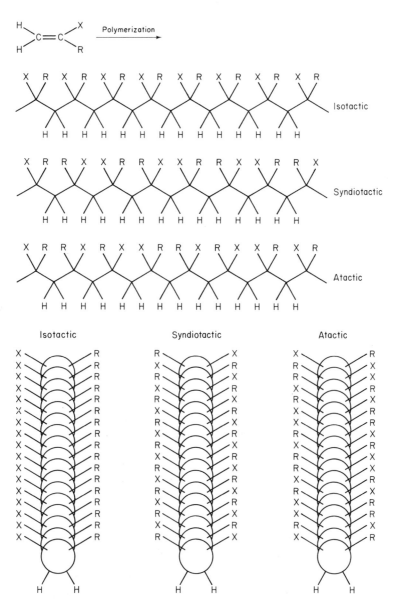

Figure 1.9. The tactic forms of vinyl polymer chains.

Polymerization of dienes can also produce polymers in which the carbon moieties are on the same side of the newly formed double bonds (cis) or on the opposite side (trans). The cis isomer is a soft elastomer; the T_g of the cis isomer of poly-1,4-butadiene is -108 °C. The trans isomer of poly-1,4-butadiene is hard, with a T_g of -83 °C.

Unsymmetrical 1,4-diene monomers such as isoprene and chloroprene can also undergo both 1,2 addition and 3,4 addition, and each of these structural isomers can be atactic, isotactic, or syndiotactic, giving eight different (possible) isomeric forms (Figure 1.10).

The *trans*-poly-1,4-butadiene isomer is a harder and less soluble rigid crystalline polymer than the cis isomer. As shown by the skeletal structures for the trans isomer (Figure 1.11), chain extensions on opposite sides of the double bonds allow good fitting of adjacent polymer chains, and this results in a rigid structure. In contrast, the *cis*-poly-1,4-butadiene isomeric polymer units do not permit such interlocking of alternate units. Even so, chain

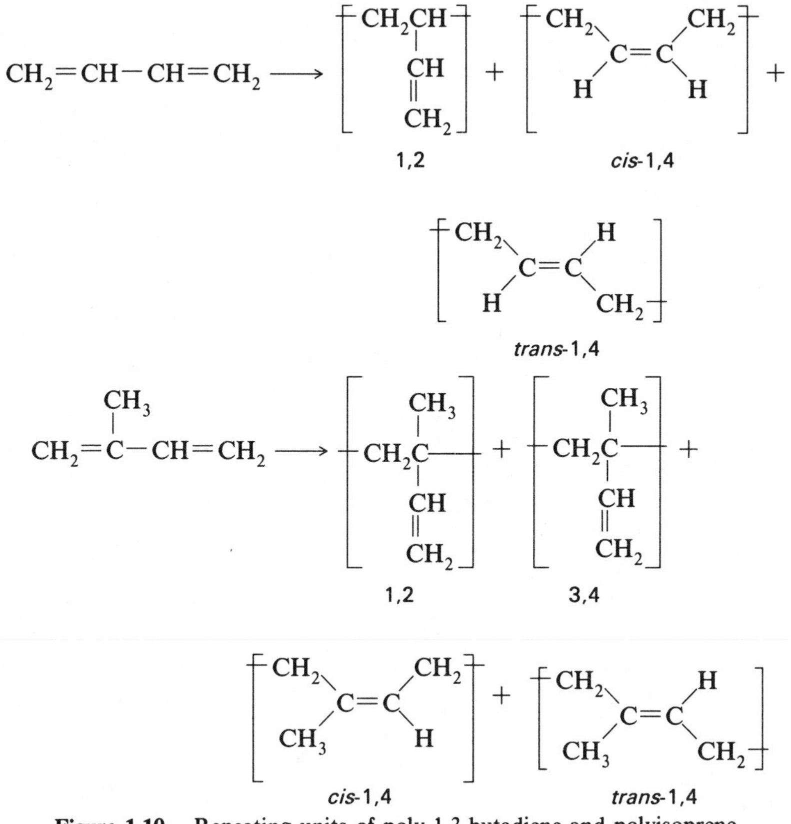

Figure 1.10. Repeating units of poly-1,3-butadiene and polyisoprene.

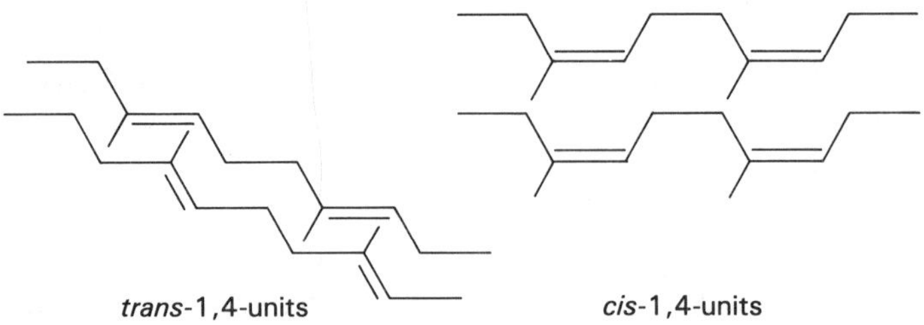

trans-1,4-units   cis-1,4-units

Figure 1.11. Skeletal representations of units of 1,4-dienes derived from 1,4-buta-diene, chloroprene, and isoprene, and other 1,4-dienes.

alignment and crystallization take place as a result of the uncoiling of randomly oriented chains during the reversible stretching process.

Natural rubber (*Hevea brasiliensis*) is *cis*-poly-2-methyl-1,4-butadiene, and gutta-percha (*Palaquium oblongifolium*) and balata (*Minusops globosa*) are polymers of isoprene (2-methyl-1,4-butadiene) with trans configurations. Neoprene is a polymer of 2-chloro-1,3-butadiene (chloroprene).

1.3 Copolymers

All polymers discussed so far are homopolymers, i.e., they consist of multiple sequences of the same repeating unit. Regular linear homopolymers without bulky pendant groups, such as HDPE, are easily crystallized. However, the tendency for crystallization is reduced in copolymers, since they contain more than one repeating unit in the chain. Copolymers with random arrangements of repeating units in the polymer chain are generally amorphous.

Although HDPE and *it*-PP are crystalline, the commercial random copolymer of ethylene and propylene (EP) is an amorphous elastomer. The most widely used EP copolymer (EPDM) is produced by the copolymerization of ethylene and propylene with a small amount of an alkyldiene; this permits cross-linking or vulcanization.

Other commercial copolymers which are typically random are those of vinyl chloride and vinyl acetate (Vinylite), isobutylene and isoprene (butyl rubber), styrene and butadiene (SBR), and acrylonitrile and butadiene (NBR). The accepted nomenclature is illustrated by EP, which is designated poly-ethylene-co-propylene the *co* designating that the polymer is a copolymer. When the copolymers are arranged in a regular sequence in the chains, i.e., ABAB, the copolymer is called an *alternating copolymer*. A copolymer consisting of styrene and maleic anhydride (SMA) is a typical alternating copolymer.

The term *block copolymer* is used to describe copolymers with long sequences or runs of different monomers in the same continuous chain, i.e., $+(A_nB_m)$. Block copolymers of styrene and butadiene (Solprene) and styrene–butadiene (SB) and styrene (S; forming polymers as $(SB)_n(S)_m$; Kraton) are commercially available. The accepted nomenclature for the former is polystyrene-b-butadiene, the *b* inserted to signify the block nature of the product.

The properties of block copolymers are dependent on the length of the sequences of repeating units, or domains. The domains in typical commercial block copolymers of styrene and butadiene are sufficiently long such that the products are flexible plastics. They are called thermoplastic elastomers (TPE). It should be noted that although the T_g for random copolymers is between the T_g's of the respective homopolymers, the repeating sequences in block copolymers exhibit their own characteristic T_g's.

The term *graft copolymer* is used to describe copolymers with long sequences of another monomer (comonomer) as branches on the main polymer chain. Most commercial varieties of high-impact polystyrene (HIP) and copolymers of acrylonitrile, butadiene, and styrene (ABS) are graft copolymers in which the main polymer chain is polybutadiene and the branches are styrene, or styrene and acrylonitrile. Figure 1.12 shows various types of copolymers.

1.4 Heteroatomic Polymers

The chains in the polymers discussed so far consist of isochains of catenated carbon atoms. However, many polymers such as the polyacetal—polyoxymethylene (POM), Delrin—have other atoms in addition to the carbon atoms in the polymer chain. As shown by the abbreviated segmental formula for the polyacetals

$$+ CR_2 - O +_n$$

they are heteropolymers with heterochains containing carbon and oxygen atoms.

The oxygen atoms in polyethylene oxide, POM, enhance the free rotation of the atoms in the chain. Nevertheless, because of good structural symmetry, these polymers are crystalline.

For reasons of simplicity, the polymers discussed so far are related to polyethylene and are usually produced by a chain reaction polymerization of vinyl monomers. Like other chain reactions, the polymerization requires three steps, i.e., initiation, propagation, and termination. Additional information

Linear - Random

$$-A-A-B-A-B-B-A-A-B-A-B-$$

Linear - Alternating

$$-A-B-A-B-A-B-A-B-$$

Linear - Block

$$-A-A-A-A-B-B-B-B-B-B-A-A-A-A-B-B-$$

Graft

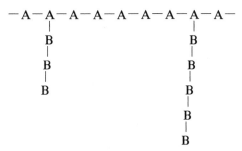

Figure 1.12. Types of copolymers.

on polymerization is not provided in this book but may be found in any introductory polymer science textbook, such as those listed at the end of this chapter.[2-10]

A large percentage of the total volume of commercial polymers—many of which are amorphous—is produced by chain or addition polymerization of vinyl monomers. Nevertheless, several important widely used commercial polymers and most of the engineering plastics and fibers are produced by stepwise or condensation polymerization of two reactants. Many of these engineering plastics and most fibers are crystalline products.

Although these more complex polymers are discussed separately, it is important to note that the statements made previously about the relation of properties to structure apply to all polymers regardless of the method of synthesis.

The first commercial synthetic linear step reaction or condensation polymer was a polyamide (nylon 66) synthesized by Carothers in the 1930's.

The repeating units in nylon 66 and in most step reaction polymers consist of residues from both reactants. The following equation shows the synthesis of nylon 66; in this case the reactants are a dicarboxylic acid (adipic acid) and a diamine (1,6-hexanediamine).

$$n\,HO_2C(CH_2)_4CO_2H \; + \; n\,H_2N(CH_2)_6NH_2 \longrightarrow$$

$$[^-O_2C(CH_2)_4CO_2^-][H_3\overset{+}{N}(CH_2)_6\overset{+}{N}H_3]_n \xrightarrow{\text{heat}}$$

$$\left[\begin{array}{c} O \quad\quad O \\ \parallel \quad\quad \parallel \\ C(CH_2)_4CNH(CH_2)_6NH \end{array} \right]_n + 2n\,H_2O$$

Adipic acid $+$ 1,6-hexanediamine $\xrightarrow{\text{heat}}$ nylon 66 $+$ water

Nylon 66 is a linear heteropolymer which, because of its symmetry, is crystalline and relatively high-melting. Many different nylons can be produced by using homologues of adipic acid and hexamethylenediamine. Polymers with pendant groups can be produced by using isomeric reactants or by reacting the amide groups with formaldehyde or ethylene oxide. These polymers with pendant groups tend to be more flexible and have lower melting points than nylon 66. Aromatic nylons (aramids) are high-melting stiff polymers because of the presence of the rigid phenylene group.

Linear polyesters, such as polyethylene terephthalate (PET, Dacron, Mylar), are heteropolymers with carbon and oxygen atoms in the polymer chain:

$$-OCH_2CH_2-O\left[\begin{array}{c} O \\ \parallel \\ C \end{array} - \bigcirc - \begin{array}{c} O \\ \parallel \\ C \end{array} -OCH_2CH_2-O \right]_n \begin{array}{c} O \\ \parallel \\ C \end{array} - \bigcirc - \begin{array}{c} O \\ \parallel \\ C \end{array} -$$

PET is a stiff crystalline polymer with a relatively high melting point. Aliphatic polyesters with more methylene groups than PET in the chain are more flexible and less crystalline than PET.

Polyurethanes (PUs) are produced by the reaction of a diisocyanate and a diol or higher polyol. As in the case of nylons and polyesters, the aliphatic PUs and those with many methylene groups between functional groups can be more flexible than aromatic PUs. A typical polyurethane is synthesized using the reactants 1,4-butanediol and hexamethylene diisocyanate as shown below:

nHO(CH$_2$)$_4$OH + nOCN(CH$_2$)$_6$NCO $\longrightarrow$

                    Hexamethylene

1,4-Butanediol   diisocyanate

$$\left[-O-(CH_2)_4O-\underset{\underset{O}{\|}}{C}-\overset{\overset{H}{|}}{N}(CH_2)_6\overset{\overset{H}{|}}{N}-\underset{\underset{O}{\|}}{C}- \right]_n$$

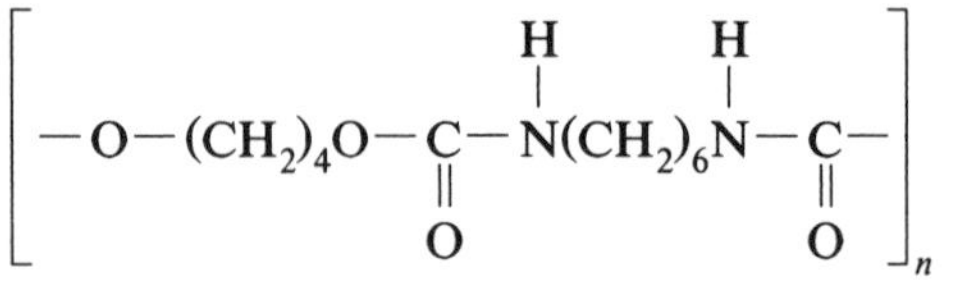

Polyurethane

Proteins, such as wool or silk, are naturally occurring linear polyamides consisting of many different amino acid repeating units. Cellulose and starch are naturally occurring linear carbohydrate polymers consisting of D-glucose repeating units joined by oxygen atoms. The repeating units in starch are joined by alpha acetal linkages like those in maltose, whereas those in cellulose are joined by beta acetal linkages like those in cellobiose. The repeating units in cellulose are shown in Figure 1.13.

As a result of the equatorial (i.e., beta) linkages between the glucose units in cellulose, cellulose is a rigid, highly crystalline polymer, is insoluble in water, and chars before melting. Derivatives of cellulose, such as esters and ethers, are less crystalline and more readily soluble in selected solvents. Because of the alpha linkages between the glucose repeating units in amylose (a starch), this polymer is more flexible than starch and is soluble in water.

Phenolic, epoxy, urea, melamine, and polyester (alkyd) polymers are cross-linked (thermoset) plastics. They are solvent-resistant and are not softened by heat. Unlike the thermoplastic step reaction polymers, which are produced by the condensation of two difunctional reactants, these network polymers are produced from reactants at least one of which has a degree of functionality higher than two.

Phenol, urea, and melamine act as, respectively, trifunctional, tetrafunc-

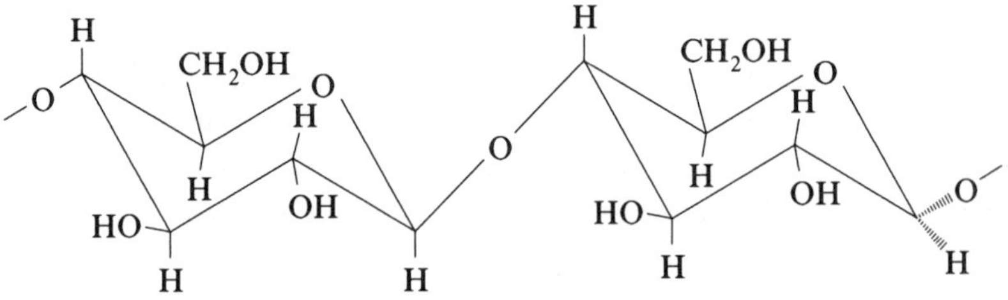

Figure 1.13. Portion of a cellulosic chain.

tional, and hexafunctional reactants. Polymers are produced by reacting these multifunctional reactants with difunctional formaldehyde. Thermoset polyesters are produced by using glycerol or an unsaturated carboxylic acid such as maleic acid as a reactant. Epoxy resin prepolymers are linear polymers which are produced by the reaction of bisphenol A and epichlorohydrin. They are cross-linked by polyfunctional amines at moderate temperatures or by cyclic anhydrides at elevated temperatures. Figure 1.14 shows a cross-linked polymer.

Nylon, polyacetal, polycarbonates, poly(2,6-dimethyl)phenylene oxide (PPO), polyimides, polyphenylene sulfide (PPS), polyphenylene sulfones, polyaryl sulfones, polyalkylene phthalates, and polyarylether ketones (PEEK) are stiff high-melting polymers which are classified as engineering plastics. The formulas for the repeating units of some of these engineering plastics are shown in Figure 1.15.

In addition to the heteropolymers having carbon and other atoms in the polymer chain, there are also several polymers, such as the silicones and phosphazenes, which have other atoms instead of carbon in the chain. As shown by the abbreviated formulas for silicone (left) and phosphazene (right),

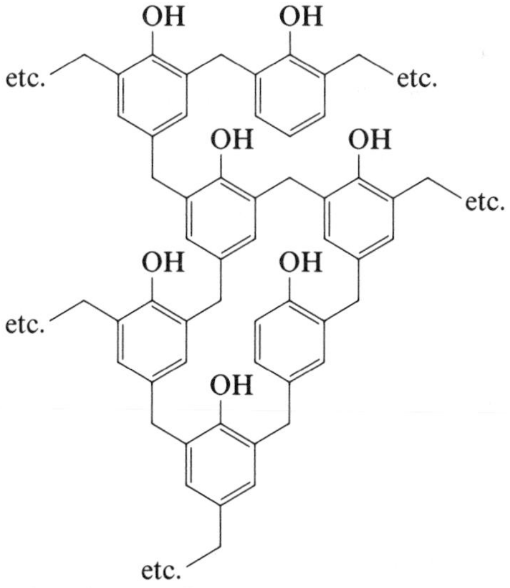

Figure 1.14. Portion of a cross-linked phenol–formaldehyde polymer.

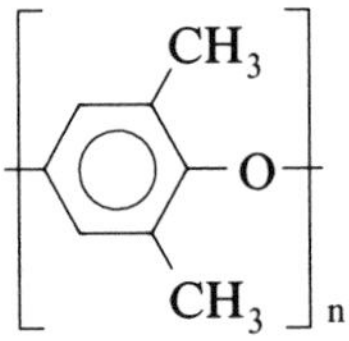

Polyphenylene oxide (PPO) 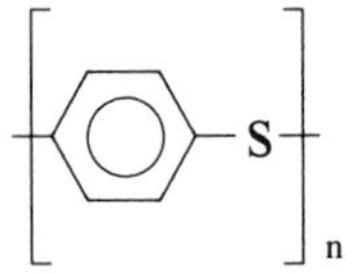   Polyphenylene sulfide (PPS)

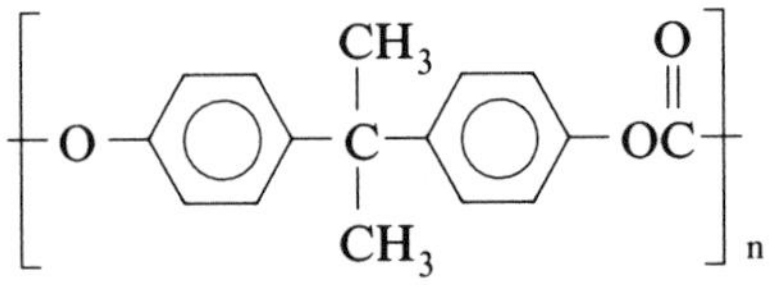

Polycarbonate (PC) 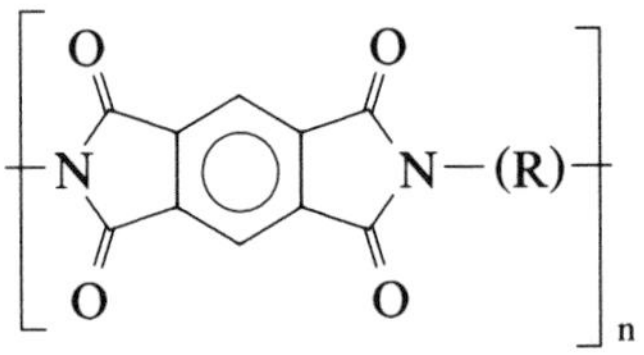   Polyimide (polymellitimide) (PI)

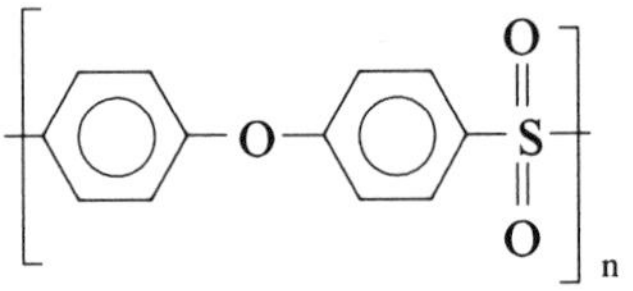

Polyaryl sulfone 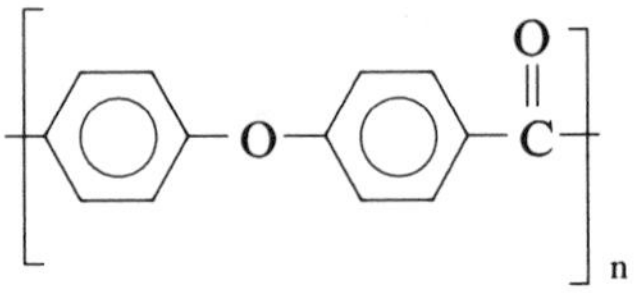   Polyarylether ketone (PEEK)*

Figure 1.15.   Repeating units of some common engineering plastics.

the repeating units in the backbone of silicones (siloxanes) contain silicon and oxygen atoms. The backbone of the phosphazenes contains phosphorus and nitrogen atoms. Silicones and phosphazenes are amorphous semi-inorganic polymers which are stable at elevated temperatures. Many of the properties of these polymers, such as lubricity, are dependent on the type of pendant group present. The thermal stability is related, to a large extent, to the bond strength of the atoms in the backbone.

1.5   References

1. H. Staudinger, *Die Hochmolecularen*, Springer-Verlag, Berlin (1932).
2. F. W. Billmeyer, *Textbook of Polymer Science*, Wiley-Interscience, New York (1971).
3. J. M. G. Cowie, *Polymer Chemistry and Physics of Modern Materials*, Intext, New York (1974).

* Industrially this is often called a polyetherether ketone to emphasize the ether character; hence the abbreviation PEEK. In actuality it is a polyarylether ketone.

4. P. J. Flory, *Principles of Polymer Chemistry*, Cornell University Press, Ithaca, N. Y. (1953).

5. F. Rodriguez, *Principles of Polymer Systems*, McGraw-Hill, New York (1971).

6. R. B. Seymour, *Modern Plastics Technology*, Reston Publishing, Reston, Va. (1975).

7. R. B. Seymour and C. E. Carraher, *Polymer Chemistry: An Introduction*, Dekker, New York (1981).

8. M. P. Stevens, *Polymer Chemistry*, Addison-Wesley, Reading, Mass. (1975).

9. H. R. Allcock and F. W. Lampe, *Contemporary Polymer Chemistry*, Prentice-Hall, Englewood Cliffs N.J. (1981).

10. H. S. Kaufman and J. J. Falcetta, *Introduction to Polymer Science & Technology*, Wiley-Interscience, New York (1977).

2 | Physical Structure of Polymers

2.1 Introduction

The molecular weight or molar mass of proteins and nucleic acids (DNA, RNA) is identical for each specific species, e.g., the molecular weights of all casein molecules from a specific source are identical. These polymers are members of a homologous series and are said to be *monodisperse* or *molecularly homogeneous.*

In contrast, other natural polymers, such as natural rubber (NR), cellulose, and most synthetic polymers, consist of mixtures of many members of a homologous series. These homologues contain varying numbers of units and thus have different molecular weights, and hence the polymers are said to be *polydisperse* or *polymolecular.*

Since most commercial polymers are polydisperse and this disparity influences the properties of the polymer, one is concerned with the molecular weight distribution and the average molecular weight $\overline{M}$. This value may be represented by an arithmetic average or number average molecular weight $\overline{M}_n$ or by a weight average molecular weight $\overline{M}_w$, which is biased toward large molecules.

2.2 Melt Viscosity

It is important to note that there is a critical threshold molecular weight below which there is little if any entanglement of polymer chains. Melt viscosity is a measure of the tendency (speed) of melted materials to flow.

The melt viscosity of a polymer increases as the molecular weight increases and is proportional to the molecular weight up to the critical threshold molecular weight. The melt index is a measure of flow inversely related to melt viscosity and is often cited in particular terms such as the time for ten grams of a material to pass through a standard orifice at a specified time and temperature. Above this value, the melt viscosity is related to the molecular weight raised to the 3.4th power. The critical threshold chain lengths corresponding to the threshold molecular weights of polymethyl methacrylate (PMMA), polyisobutylene, and polystyrene (PS) are 208, 610, and 730 repeating units, respectively.

Most industrial polymers have average chain lengths above critical threshold value. In general, the physical properties improve rapidly as the threshold value is approached and then tend to level off above this value.

2.3   Interchain and Intrachain Forces

The forces present in molecules are often divided into primary forces (typically greater than 50 kcal/mol) and secondary forces (typically less than 10 kcal/mol). Primary bonding forces can be subdivided into ionic (between atoms of greatly differing electronegativities; characterized by a lack of directional bonding; not typically present within polymer backbones), metallic (the number of outer, valence electrons is too small to fill complete outer shells; metallically bonded atoms are often considered as charged atoms surrounded by a potentially fluid sea of electrons; lack of bonding direction; not found in polymers), and covalent (including coordinate and dative; the major means of bonding within polymers; directional) bonding. The bonding lengths of primary bonds are usually about 0.09 to 0.20 nm; the carbon–carbon bond length is about 0.15 to 0.16 nm.

Secondary forces, frequently called *van der Waals forces* since they are the forces responsible for the van der Waals correction to the ideal gas relationships, interact over longer distances than primary forces, generally having significant interaction between .25 and .50 nm. The force of these interactions is inversely proportional to some power of r, generally 2 or greater [force $\propto 1/(\text{distance})^r$], and thus is dependent on the distance between the interacting molecules. Thus many physical properties of polymers are dependent on both the conformation (arrangements related to rotation about single bonds) and the configuration (arrangements related to the actual chemical bonding about a given atom) since both affect the proximity of one chain to another.

Atoms in individual polymer molecules are joined to each other by relatively strong covalent bonds. The bond energies of the carbon–carbon bonds are on the order of 80 to 90 kcal/mol. Polymer molecules, like all other molecules, are attracted to each other by intermolecular secondary forces.

Intermolecular forces are responsible for the increases in boiling points within a homologous series such as the alkanes, for the higher-than-expected boiling points of polar organic molecules such as alkyl chlorides, and for the abnormally high boiling points of alcohols, amines, and amides. Although the forces responsible for these increases in boiling points are all called van der Waals forces, these forces are subclassified in accordance with their source and intensity. Secondary, intermolecular forces include *London dispersion forces, induced polar-dipolar forces,* and *dipolar forces,* including *hydrogen bonding.*

Nonpolar molecules such as ethane $H(CH_2CH_2)H$ and polyethylene $(CH_2CH_2)_n$ are attracted to each other by weak *London* or *dispersion forces* resulting from induced dipole–dipole interactions. The temporary dipoles in ethane or along the polyethylene chain are due to instantaneous fluctuations in the density of the electron clouds caused by constant motion of electrons about the nucleus with the homogeneity upset by similar electron movement about the other nucleus. The energy of these forces is about 2 kcal per mole of repeating unit in nonpolar and polar polymers alike, and this force is independent of temperature. These dispersion forces are the major forces present between chains in many elastomers and soft plastics.

$$\begin{array}{cc} \ce{>C=O\cdots H-\underset{\displaystyle H}{\overset{\displaystyle H}{C}}-H} & \ce{>C=O\cdots >C=O} \end{array}$$

induced polar	dipolar
directional	directional

It is of interest to note that under normal conditions methane, ethane, and ethene are all gases, and hexane, octane, and nonane are all liquids, whereas polyethylene is a waxy solid. This trend is primarily due to both an increase in mass per molecule and an increase in the London forces per molecule as the chain length increases.

Polar molecules, such as ethyl chloride, $H_3C\text{-}CH_2Cl$, and polyvinyl chloride (PVC), $(CH_2\text{-}CHCl)_n$, are attracted to each other by *dipole–dipole interactions* resulting from the electrostatic attraction of a chlorine atom in one molecule to a hydrogen atom in another molecule. These dipole–dipole interactions, whose energies range from 2 to 6 kcal per mole of repeating unit in the molecule, are temperature-dependent, and these forces are reduced as the temperature is increased in the processing of polymers. Although the dispersion forces are typically weaker than the dipole–dipole forces, they are also present in polar compounds.

Strongly polar molecules, such as ethanol, polyvinyl alcohol, and cellulose, are attracted to each other by a special type of dipole–dipole interaction called *hydrogen bonding* in which the oxygen atoms in one molecule are attracted to the hydrogen atoms in another molecule. These are the strongest of the intermolecular forces and may have energies as high as 10 kcal per mole of repeating unit (the energy of the HF hydrogen bond is higher). Intermolecular hydrogen bonds are usually present in fibers, such as cotton, wool, silk, nylon, polyacrylonitrile (PAN), polyesters, and polyurethanes (PUs). Intramolecular hydrogen bonds are responsible for the helices observed in starch and globular proteins.

It is important to note that the high melting point of nylon 66 (265°C) is the result of a combination of dispersion, dipole–dipole, and hydrogen bonding forces between the polyamide chains. The number of hydrogen bonds is decreased when the hydrogen atoms in the amide groups in nylon are replaced by methyl groups and when the hydroxyl groups in cellulose are esterified or etherified.

The principal difference in the physical properties of polyethylene and paraffin wax is based on both chain entanglement and total intermolecular dispersion forces per molecule (chain).

Secondary forces operate at longer distances (.25–.50 nm) than covalent bonds. These secondary forces are much weaker than primary covalent bonds, but the forces are cumulative. Thus the cohesive energy of a polymer is equal to the summation of the cohesive energy density (CED) values for each mole of repeating unit (< 2 kcal) in the chain. The CED of a liquid is defined as the energy of vaporization per unit volume, $\triangle E/V$.

In addition to the strength resulting from chain entanglement and cumulative intermolecular dispersion forces, more-polar polymers, such as PVC and PMMA, have cumulative dipole–dipole intermolecular forces which are about 5 kcal for each repeating unit. Likewise, more-polar polymers, such as polyamides and cellulose, also have strong cumulative intermolecular hydrogen bonds whose energies are as large as 10 kcal for each mole of repeating unit.

The secondary intermolecular force, or CED, is derived from the intermolecular energy (including dispersion, hydrogen bonding, and polar components) of the repeating unit in the polymer. This value varies from 36 cal · cm^{-3} for polytetrafluoroethylene (PTFE) to 231 cal · cm^{-3} for PAN. In fact the low CED of PTFE is largely responsible for its use as a nonstick coating.

2.4 Glass Transition Temperature

Because of the kinetic energy present in the molecule, amorphous flexible polymer chains are usually in constant motion at ordinary temperatures. The extent of this wiggling-like segmental motion decreases as the temperature is lowered in this reversible process. The temperature at which this segmental or micro-Brownian motion of amorphous polymers becomes significant as the temperature is increased is called the glass transition temperature, T_g. The term *free volume* is used to describe the total volume occupied by the holes.

The change in free volume or vacant volume of a polymer as temperature is increased is small below the T_g, but the rate of change increases abruptly at the T_g. In fact, measuring the change in the slope of the curve plotting free volume as a function of temperature is one method of obtaining the T_g. The free volume is largely independent of the type of polymer and may be as high as 10% of the total volume at the T_g. The free volume of most polymers at their T_g's is similar. Thus the free volume of most polymers is similar below and at the T_g.

Both kinetic and thermodynamic approaches have been used to measure and explain the abrupt change in properties as a polymer changes from a glassy to a leathery state. These involve the coefficient of expansion, the compressibility, the index of refraction, and the specific heat values. In the thermodynamic approach used by Gibbs and DiMarzio, the process is considered to be related to conformational entropy changes with temperature and is related to a second-order transition. There is also an abrupt change from the solid crystalline to the liquid state at the first-order transition or melting point T_m.

Although the transition from a solid to a liquid state at the T_m is relatively precise and occurs over a short temperature range, the transition from a glassy solid to a leathery state occurs over somewhat broader temperature ranges around T_g. The modulus, or stiffness, of the polymer decreases as the temperature is increased above the T_g, and the polymer changes from a leathery to a rubbery state.

While in the temperature range called the rubbery plateau, the soft polymer responds instantaneously and reversibly to applied stress and tends to be Hookean. In the rubber state, the polymer approaches Hooke's law for ideal elastic solids, i.e., the stress S is proportional to the strain γ, i.e., $S = G\gamma$. The proportionality constant is the modulus G. Most commercial elastomers (rubbers) are used within the rubbery plateau.

It is possible that there are in fact two secondary types of transitions: one, the transition between the glassy and leathery states, called the T_g; and an additional transition occurring between the leathery and rubbery regions.

As the temperature is increased above the rubbery plateau, the linear amorphous polymer assumes a viscous state and may undergo irreversible flow, i.e., flows such that the original shape is lost. The flow of the viscous liquid may approach a Newtonian flow, i.e., its flow properties may be estimated from Newton's law for ideal liquids.

According to Newton's law, the stress S is proportional to the viscosity gradient or flow $d\gamma/dt$. The proportionality constant is the coefficient of viscosity often referred to as simple viscosity η, so $S = \eta(d\gamma/dt)$.

The value of the T_m is almost always at least 33% higher than the T_g expressed in degrees Kelvin. Symmetrical polymers, like linear polyethylene, exhibit the greatest percentage difference between the T_m and the T_g, and these values are usually well-known in symmetrical polymers.

The viscosity η of the molten polymer decreases as the temperature is increased, and this change in viscosity may be estimated from the Arrhenius equation; $\eta = Ae^{E_a/RT}$, where E_a is the activation energy required for the molecule or segment to "jump" into a hole in the solvent causing flow. The flow of polymers requires cooperative segmental motion for segments greater than 15 to 30 atoms so that the molecules can jump into the holes. The flow of polymer molecules is retarded by branches, pendant groups, and high CED values.

The cooperative segmental motion in polymer molecules can be considered as a crankshaft motion of six atoms in the polymer chain. According to H. Eyring, the viscosity of a polymer melt decreases exponentially in accordance with the enthalpy of activation ΔH_a instead of the energy of activation E_a as stated in the Arrhenius equation.

In any case, the Arrhenius equation is not particularly useful at temperatures above $T_g + 100$ K. The overall temperature-dependence of polymer flexibility at temperatures of T_g to $T_g + 100$ K can be expressed by the empirical Williams, Landel, and Ferry (WLF) equation

$$\text{Log } \eta \propto \frac{-C_1(T - T_g)}{C_2 + (T - T_g)}$$

in which the constants C_1 and C_2 are related to holes and are characteristic for each polymer system.

The viscosity of the polymer increases rapidly as the temperature is lowered toward T_g, and the polymer chains exist in many conformations as compact coils. The polymer undergoes reversible stretching when subjected to instantaneous stress in the rubbery plateau as it goes from a low to a high entropy value.

However, if this stress is maintained for long periods of time, the compact coils may unravel, permitting the chains to slip past each other forming new coils in a process called *stress decay*. Stress decay is prevented if a few cross-links are present, as in vulcanized rubber.

Stress on optically transparent polymers below the T_g may cause stretching of the fixed covalent bonds and distortion of the bond angles in the polymer chain. The behavior of the elastic solid below the T_g will approach Hooke's law for short strain or elongation, but covalent bonds will break if sufficient stress is applied.

The T_g is related to chain stiffness and the geometry of the polymer chain. Flexible polymers with methylene and oxygen atoms in the chain, such as polyethylene, polyoxymethylene, and polysiloxane (silicone), have relatively low T_g values. The T_g of polyoxymethylene is somewhat higher than would be anticipated because of the dipole character of the C—O—C group, which increases the intermolecular forces and restricts segmental motion.

Stiff polymers, such as polyphenylene, nylon 66, polyphenylene sulfone, and polyarylether ketone (PEEK), have relatively high T_g values because of the presence of phenylene and sulfone or carbonyl stiffening groups in the chain.

Bulky pendant groups also restrict segmental motion, and this effect on T_g values is enhanced when the pendant groups are polar. Thus polypropylene (PP) ($T_g = 253$ K) has a higher T_g than polyethylene ($T_g = 147$ K), and polyvinyl alcohol (PVA) ($T_g = 358$ K) has a higher T_g than either of these polyolefins.

Since alpha substituents also impede segmental motion, PMMA ($T_g = 378$ K) has a higher T_g than polymethyl acrylate (PMA) ($T_g = 279$ K). The various factors which affect T_g values are additive.

The T_g values of polymers with flexible pendant groups decrease as the size of the pendant groups increases. Thus the T_g values for polymers of ethyl, propyl, and butyl acrylate are 249, 225, and 218 K, respectively.

However, there is an optimum size for pendant groups with respect to decreasing T_g. Because of intramolecular attraction between large pendant groups, the T_g is increased when the ester groups have more than 10 carbon atoms. This effect is called *side chain crystallization*.

The polymer *cis*-1,4-polybutadiene ($T_g = 170$ K) is more flexible and has a lower T_g than *trans*-1,4-polybutadiene ($T_g = 190$ K), and in general this is true of all geometric isomers.

The T_g values of cross-linked polymers are similar to those of comparable linear polymers providing the cross-linked polymer has a low cross-linked density, i.e., 30 to 50 repeating units between the cross-links. The T_g values increase as the cross-link density increases. Thus the T_g of hard rubber which is more highly cross-linked is higher than that of soft vulcanized rubber.

The T_g values may also be reduced by random copolymerization with a monomer which produces more-flexible polymers. This lowering of T_g values by copolymerization is sometimes called *internal plasticization*.

In general, the T_g of a random polymer may be estimated from the law of mixtures. For example, the T_g of a copolymer containing equimolar quantities of styrene and 2,4-dimethylstyrene is 381 K, while the T_g values of polystyrene and poly-2,4-dimethylstyrene are 356 and 404 K, respectively. The T_g of random copolymers may be estimated from

$$\frac{1}{T_g} = \frac{W_1}{T_{g^1}} + \frac{W_2}{T_{g^2}}$$

which is the equation for the T_g of random copolymer M_1M_2, where W_1 and W_2 are weight fractions of the respective co-units in the copolymer.

It is important to note that the T_g values of component polymers may be unaffected when they are present in multiphase blends (separate phases at the microlevel), and this is the basis for many multiviscosity oils. This is also true for block and graft copolymers, which have characteristic T_g values corresponding to the polymers of each of the comonomers.

The T_g value is reduced by the addition of moderate amounts of plasticizers (additive which reduces intermolecular forces) and is sometimes even increased by the addition of small amounts of plasticizers (antiplasticization) and optimum amounts of fillers (usually a relatively inert material used as the discontinuous phase of a composite) and reinforcements (materials such as fibrous additives which give increased strength to a polymer).

2.5 Crystallinity

The chemical structure determines a given polymer's tendency toward being crystalline or amorphous, or being a mixture of crystalline and amorphous regions in the solid state. In general, crystallinity is favored by symmetrical chain structures that allow close packing of the polymer units to take greater advantage of secondary forces which are distance related. Crystallinity is also favored by high interchain (secondary forces) interactions. Thus linear polyethylene (HDPE) and PTFE, both with highly symmetrical chain structures, have a high tendency to form crystalline solids. Atactic PVC, with its asymmetrical chlorine atom, tends toward being amorphous. Polyvinylidene chloride (PVDC), with its two chlorine atoms, tends to be crystalline, since the chlorine atoms are symmetrical. Atactic PVA is partially crystalline because of the presence of specific interchain interactions, in this case hydrogen bonding.

Factors effecting increased hydrogen bonding also bring about (in general) increases in many properties including T_g and T_m. HDPE, although highly symmetrical, has a low T_m (135 °C), whereas nylon 66, both symmetrical and possessing good hydrogen bonding, exhibits a T_m of about 270 °C.

Single crystals with a T_m of 423 K have been produced from low-density polyethylene (LDPE). Isotactic PP crystals have a T_m of 444 K and syndiotactic PP has a T_m of 411 K, whereas atactic PP is amorphous and has a T_g of 255 K.* Isotactic polyolefins with pendant groups, such as polyhexene, have high T_m values. Random copolymers of ethylene and propylene are amorphous, but block copolymers of these monomers are crystalline.

The effect of structural regularity on properties of polymers may be illustrated by the hydrolytic products of polyvinyl acetate (PVAc). PVAc is insoluble in water, but because of the presence of polar hydroxyl groups, partially hydrolyzed PVAc is soluble in water.

Tacticity and geometric isomerism affect the tendency toward crystallization: the tendency increases as the tacticity (stereoregularity) is increased and when the geometric isomers are predominantly trans. Thus isotactic PS is crystalline, whereas atactic PS is largely amorphous; and cis-polyisoprene is amorphous, whereas the more easily packed trans isomer is crystalline.

Although most unstretched elastomers (cf,2,6) and many plastics (cf,2,8) are amorphous, most fibers are highly crystalline.

Crystallizable polymers tend to form randomly oriented crystallites which are oriented when the polymer is stretched or cold drawn at temperatures below the T_m. Crystallization under pressure may result in a fibrillar structure or extended chain structure.

Crystalline polymers are less soluble below the T_m than amorphous

---

* Highly crystalline polymers have no T_g, only a T_m, whereas highly amorphous polymers generally exhibit only a T_g.

polymers. When a crystallizable polymer is cooled to a temperature below T_m, it may form clusters of disoriented crystallites called *spherulites.*

The density of a polymer is related directly and the transparency is related inversely to the degree of crystallinity.

The degree of crystallinity may be calculated from the density of the polymer if the density is known for the amorphous and crystalline states. Some crystallizable polymers are polymorphic, i.e., they may exist in more than one crystalline form. An unstable crystalline form may change to a more stable form, and crystalline forms may change under stress. For example, HDPE changes from an orthorhombic crystalline polymer to a monoclinic form when subjected to compressive forces.

Although it is called *crystal clear polystyrene,* atactic PS is amorphous and clear, as are PMMA and polycarbonate (PC). In contrast, HDPE and nylon 66 are highly crystalline and opaque. The relationship of packing efficiency to specific gravity and crystallinity is readily illustrated by LDPE and HDPE, which are 60 and 95% crystalline, respectively, and have specific gravities of 0.91 and 0.97, respectively.

Isotactic PP, which is a commercial variety of PP, exists as a helix instead of an extended planar conformation in order to relieve the strain and thus attain a state of low free energy. Such helical conformations are actually rodlike and thus pack parallel to other rods in the crystal lattice.

Most elastomers are amorphous, but those with regular structures can crystallize when cooled to extremely low temperatures. Vulcanized soft rubber, which has a low cross-link density, when stretched crystallizes in a reversible process, and the oriented polymer has a high modulus (high stress for small strains, i.e., stiffness) and high tensile strength.

Stretched elastomers and oriented fibers are characterized by a high degree of alignment and are highly crystalline. When stretched, isotactic PP can be used as a crystalline fiber, though the amount of secondary bonding is small. Most other fibers are characterized by strong intermolecular hydrogen bonding. However, it is important to note that in addition to crystalline domains, all polymers have some amorphous domains, even if they are simply the intersection of chain folds and end groups.

Nylon 66 (Figure 2.1) and polyethylene terephthalate (PET) are high-melting fibers which exist as intermolecularly hydrogen bonded zigzag chains which crystallize because of the symmetry present in the polymer chains and the hydrogen bonding occurring between chain segments. Similar arrangements are characteristic of acrylic (PAN), PU and cellulosic fibers. In contrast, some polymers, such as proteins and nucleic acids, are characterized by a

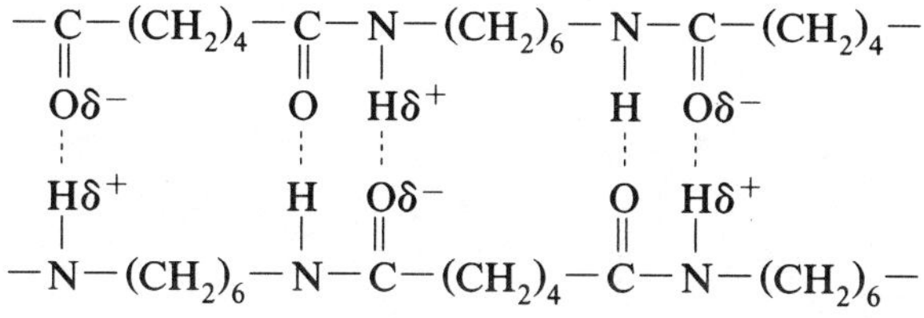

$$-\underset{\underset{\overset{\vdots}{H\delta^+}}{O\delta^-}}{\underset{\|}{C}}-(CH_2)_4-\underset{\underset{H}{\overset{\|}{O}}}{C}-\underset{\underset{O\delta^-}{H\delta^+}}{N}-(CH_2)_6-\underset{\underset{O}{H}}{N}-\underset{\underset{\overset{\vdots}{H\delta^+}}{O\delta^-}}{\underset{\|}{C}}-(CH_2)_4-$$

$$-\underset{H\delta^+}{N}-(CH_2)_6-\underset{H}{N}-\underset{O\delta^-}{\overset{\|}{C}}-(CH_2)_4-\underset{O}{\overset{\|}{C}}-\underset{H\delta^+}{N}-(CH_2)_6-$$

Figure 2.1 Hydrogen bonding between nylon 66 units.

helical conformation rather than stretched zigzag chains in the crystalline state.

When produced from reactants with an even number of carbon atoms, nylons (polyamides), PUs, and polyesters are more compact and have higher T_m values than those produced from reactants with an odd number of carbon atoms.

The T_m of condensation polymers such as polyesters and polyamides is decreased as the number of methylene (CH_2) groups in the reactants is increased. The presence of stiffening groups, such as phenylene groups in a polymer chain, increases the T_m.

The T_m is always greater than the T_g. The ratio of T_m to T_g ranges from about 1.3 for polydimethylsiloxane to about 3 for polytetrafluoroethylene.

2.6 Elastomers

Elastomers are high-molecular-weight polymers possessing chemical and/or physical cross-linking. For industrial application the temperature at which the elastomer is used must be above the T_g (to allow for full "chain" mobility), and in its normal state (unextended), the elastomer must be amorphous. The restoring force, after elongation, is largely entropic. As the material is elongated, the random chains are forced to occupy more-ordered positions. On release of the applied force, the chains tend to return to a more-random state.

Gross mobility of entire chains must be low. The cohesive energy forces between chains of elastomers permit rapid, easy expansion. In its extended state, an elastomeric chain exhibits a high tensile strength, whereas at a low extension it has a low modulus. Polymers with low cross-link density usually meet the desired property requirements. The material after deformation returns to its original shape because of the cross-linking. This property is often referred to as *elastic memory*.

2.7   Fibers

Fibers are characterized by high tensile strength and high modulus. These properties are associated with much molecular symmetry and high cohesive energies between chains, both of which are associated with a fairly high degree of polymer crystallinity. Fibers are normally linear and drawn (oriented) in one direction and thus have high mechanical properties (as tensile strength, modulus, flexural strength) in that direction.

Branching and cross-linking in fibers are undesirable since they disrupt crystal formation. However, a small amount of cross-linking may increase some physical properties such as tensile strength if incorporated in the polymer after the material is suitably drawn and processed.

2.8   Plastics

Materials with properties intermediate between those of elastomers and fibers are grouped together under the term *plastics*. Thus plastics exhibit some flexibility and hardness with varying degrees of crystallinity. The molecular requirements for a plastic are that (1) if it is linear or branched, with little or no cross-linking, it be below its T_g, (2) if it is amorphous and/or crystalline, it be used below its T_m, or (3) if it is cross-linked, the cross-linking be sufficient to severely restrict molecular motion.

2.9   References

T. Alfrey and E. F. Gurnee, *Organic Polymers*, Prentice-Hall, Englewood Cliffs, N.J. (1967).

F. Bveche, *Physical Properties of Polymers*, Wiley-Interscience, New York (1962).

H. G. Elias, *Macromolecules: Structured Properties*, Vol. 1, Plenum Press, New York (1977).

D. Fox, M. M. Labes, and A. Weisberger, *Physics and Chemistry of the Organic Solid State*, Wiley-Interscience, New York (1963).

L. Mandelkern, *An Introduction to Macromolecules*, Springer-Verlag, New York (1972).

P. Meares, *Polymers: Structure and Bulk Properties*, Van Nostrand, New York (1965).

L. E. Nielsen, *Mechanical Properties of Polymers*, Dekker, New York, (1974).

A. V. Tobolsky, *Properties and Structures of Polymers*, Wiley, New York (1960).

L. R. G. Treloar, *The Physics and Chemistry of Rubber*, Oxford Clarendon Press, London (1958).

3 | Tests for Properties of Polymers

3.1 Introduction

The frequency of failure (breakdown) of polymeric materials has decreased and will continue to decrease as polymer scientists and technologists recognize the importance of significant tests. In addition to knowing the glass transition temperature T_g and the melting point T_m, scientists must know the results of many other laboratory tests before a polymer can be recommended for a specific application.

Most tests for polymers have been developed by the American Society for Testing and Materials (ASTM) Committee D-20. In addition to these tests, which are published in frequent editions by ASTM, there are also specific tests developed by the Society of the Plastics Industry (SPI), the Society of Plastics Engineers (SPE), and other technical organizations. The variety of tests and results is most prominent in the areas of electronic and optical properties.

The major problem in testing today is designing tests which measure the desired property: tests that predict failure, rupture, etc., with confidence. There are many in-house tests, tests developed by a particular company to measure a specific property, which may relate to tests already accepted by the ASTM. However, results of these in-house tests should not be directly compared with results derived from other test methods.

A second major problem is the designing of tests that accurately predict accelerated aging related to particular properties.

The following tests related to important properties are discussed in this chapter:

1.   Thermal tests

2. Solubility tests
3. Tests for diffusion and permeability
4. Tests for physical or mechanical properties
5. Tests for optical properties
6. Flammability tests
7. Tests for electric properties
8. Tests for chemical resistance
9. Weatherability tests

3.2 Thermal Tests

Major methods involved with the generation of information about thermal property behavior of materials include thermal gravimetric analysis (TGA), differential scanning calorimetry (DSC), differential thermal analysis (DTA), torsional braid analysis (TBA), thermal mechanical analysis (TMA), and pyrolysis gas chromatography (PGC).

One of the simplest techniques is pyrolysis gas chromatography (PGC) in which the gases resulting from the pyrolysis of a polymer are analyzed by gas chromatography (GC). This technique may be used for qualitative and quantitative analysis. The latter requires calibration with known amounts of a standard polymer pyrolyzed under the same conditions as the unknown.

Following is a brief description of the most general criteria associated with the DSC modes.

DSC is a technique of nonequilibrium calorimetry in which the heat flow into or away from the polymer compared with the heat flow into or away from a reference is measured as a function of temperature or time. This is different from DTA, in which the temperature difference between a reference and a sample is measured as a function of temperature or time. Presently available DSC equipment measures the heat flow by maintaining a thermal balance between the reference and the sample by changing a current passing through the heaters under the two chambers. For instance, the heating of a sample and a reference (typically an empty sample pan) proceeds at a predetermined rate until heat is emitted or consumed by the sample. If an endothermic change takes place in the sample, the temperature of the sample will be less than that of the reference. The circuitry is programmed to maintain a constant temperature for both the reference and the sample compartment by raising the temperature of the sample to that of the reference. The current necessary to maintain a constant temperature of the sample and the reference is recorded. The area under the resulting curve is a direct measure of the heat of transition.

Possible determinations from DSC and DTA measurements include (1) the heat of transition, (2) the heat of reaction, (3) the sample purity, (4) the phase diagram, (5) the specific heat, (6) the sample identity, (7) the rate of crystallization, melting, or reaction, and (8) the activation energy.

In TGA, a sensitive balance is used to follow the weight change of a polymer as a function of time or temperature. In making both TGA and thermocalorimetric measurements, the same heating rate and flow of gas should be employed to give the most comparable thermograms. TGA and DTA can be used to determine (1) the sample purity, (2) the material identity, (3) the amount of solvent retained, (4) the reaction rate, (5) the activation energy, (6) the heat of reaction, and (7) the thermal stability.

TMA measures the mechanical responses of a polymer as a function of temperature. Typical measurements include (1) expansion properties, i.e., the expansion of a material leading to the calculation of the linear expansion coefficient; (2) tension properties, i.e., the shrinkage and expansion of a material under tensile stress; e.g., elastic modulus; (3) dilatometric properties, i.e., the volumetric expansion within a confining medium; e.g., specific volume; (4) single-fiber properties, i.e., the tensile response of a single fiber under a specific load; e.g., single-fiber modulus; and (5) compression properties, e.g., the softening or penetration under load.

Compressive, tensile, and single-fiber properties are usually measured under a specified load, and these tests yield information on softening points, modulus changes, phase transitions, and creep (cold flow) properties. For compressive measurements, a probe is positioned on the sample and loaded with a given stress. A record of the penetration of the probe into the polymer as a function of temperature is obtained. Tensile properties can be measured by attaching the fiber to two fused quartz hooks. One hook is loaded with a given stress, and elastic modulus changes are recorded by monitoring probe displacement.

In TBA the changes in tensile strength as the polymer undergoes thermal transition are measured as a function of temperature and sometimes also as a function of the applied frequency of vibration of the sample. The name TBA is derived from the fact that measurements are made on fibers which are "braided" together to give test samples connected onto vicelike attachments or hooks. As thermal transitions take place, irreversible changes, such as thermal decomposition or cross-linking, may be observed. In general, a change in T_g or a change in the shape of the curve of torsion versus temperature during repeated sweeps through the region indicates an irreversible change.

DSC, DTA, TMA, and TBA analyses are all interrelated, all signaling changes in thermal behavior as a function of heating rate or time. TGA is related to the other analyses in the assignment of phase changes associated with weight changes.

The polymer softening range, although not a specific thermodynamic property, is normally simply and readily obtainable and is a useful "use" property related to flexibility, hardness, etc. Softening ranges generally lie between the T_g and T_m of the polymer. Some polymers do not soften; they undergo a solid state decomposition before softening.

Softening ranges are dependent on the technique and procedure used to determine them. Thus information on softening ranges should be accompanied by information on the specific technique and procedure employed for the determination.

Softening range data can serve as guides to proper temperatures for melt fabrication, such as melt pressing, melt extruding, and molding. They also are related to the product's thermal stability.

The following techniques are often used for the determination of values associated with polymer softening ranges. The T_g is the temperature at which significant segmental motion begins as the temperature of an amorphous polymer is increased. The traditional technique for measuring T_g is by noting the abrupt change that occurs in the slope of the curve when the specific volume of an amorphous polymer is plotted against the temperature; the abrupt change occurs at T_g. Similar results are obtained if the coefficient of expansion, the heat content, or the index of refraction is plotted against temperature.

A penetrometer consisting of a weighted needle and depth gauge may be used to estimate T_g. In this crude test, the penetration of the weighted needle is monitored as the temperature of the polymer, in contact with the needle, is increased.

The so-called brittle point, associated with sample failure in impact tests, may be determined qualitatively using a penetrometer or a Shore durometer (an instrument used to measure resistance of a sample to penetration by a blunt needle) to measure the change in penetration hardness with temperature. Also, a thin film of a polymer may be readily folded at temperatures above T_g but may crack when folded at temperatures below T_g.

A torsional pendulum may also be used to monitor the loss in rigidity of a polymer at T_g. A braided glass fiber impregnated with the organic polymer is suspended in the torsional device, and the period of the pendulum and its damping characteristics are measured as the temperature is increased. As noted in Chapter 2, the free volume (directly proportional to the real or specific volume) increases abruptly at T_g as the temperature is increased. The change in specific volume of a polymer as a function of temperature may be determined by immersing the polymer in a nonsolvent liquid, such as mercury, and monitoring the increase in the volume of the liquid in a capillary tube as the temperature is raised. An automated dilatometer (apparatus employed for precisely measuring volumes) may be used for the determination of T_g.

The effect of heat on polymers at moderate temperatures may be demonstrated by the ball-and-ring method (ASTM-E28), the Vicat softening point test (ASTM-D1525), and the heat deflection test (ASTM-D648). In the ball-and-ring test, a polymer is placed in a ring, and both are heated in a liquid such as glycerol. The end point is the temperature at which a metal ball falls through the polymer to the bottom of the vessel.

In the Vicat test, the polymer sample is subjected to a load of 1 kg on a standard needle. The sample is immersed in a bath and heated, and the softening point is defined as the temperature at which the loaded needle penetrates to a depth of 1.0 mm.

A good applications-oriented measure of the use temperature for a material is the heat distortion or heat deflection test (HDT). The HDT is described by ASTM-D648 as the temperature at which a sample of defined dimensions ($5 \times \frac{1}{2} \times \frac{1}{8}$ (or $\frac{1}{4}$) in.) deflects under a flexural load of 66 or 264 psi placed at its center. In case of a largely amorphous polymer, the HDT temperature is typically slightly (10 to 20 °C) lower than the T_g as determined by DSC or DTA, whereas with more-crystalline polymers, it more closely correlates with the T_m. The HDT temperature is a useful indicator of the temperature limits for structural (load-supporting) applications. A loaded cantilever beam is used in another heat deflection test called the Martens method.

The T_m of crystalline polymers may be determined by observing the first-order transition (change in heat capacity value) by DTA or by DSC (ASTM-D3418). Some comparative information on thermal properties of polyolefins may be obtained from the melt index. To determine the melt index, the weight of extrudate or strand under a specified load and at a specified temperature is measured. Melt index values are inversely related to the melt viscosity.

The linear coefficient of expansion of a bar of polymer may be determined by measuring the difference in its length at two different temperatures and dividing by the temperature difference. The cubical coefficient of expansion may be measured in a dilatometer such as that used for measuring T_g.

The thermal conductivity of polymers may be determined by measuring the heat flow through a known thickness of solid polymer (ASTM-C177) or foam (ASTM-D2326) with a given temperature differential across the thickness. The effect of heat on polymers may be determined by heating the polymer sample in a circulating air oven for 4 h at a specified temperature (ASTM-D794).

3.3 Solubility Tests

The solubility parameter of a liquid is a numerical value equal to the square root of the heat of vaporization per unit volume and it is employed in predicting solubility and miscibility. The solubility parameter of polymers may be determined by measuring the extent of swelling in solvents with known solubility parameters. The polymer sample is prepared by copolymerizing the appropriate monomer with a small amount of a cross-linking agent. One may also dissolve the uncrosslinked polymer in a measured volume of a good solvent and titrate with a poorer (or nonsolvent) solvent until turbidity is noted. Since the law of mixtures applies to the solubility parameter, it is possible to blend nonsolvents forming a mixture which will dissolve the polymer. For example, if an equal molar mixture of n-octyl alcohol ($\delta = 10.3$) and 2-ethylhexanol ($\delta = 9.5$) were found to just dissolve a polymer, the polymer's solubility parameter would be about $9.9(10.3 + 9.5)/2$.

Solubility parameters of polymers may also be determined by preparing solutions of identical concentrations of polymer in a series of solvents with known solubility parameters. The solubility parameter of the polymer is equal to that of the solvent giving a solution with the highest viscosity. More precise values may be obtained by plotting the viscosity versus the solubility parameter of the solvent.

The retension of liquids by polymers may be determined by immersing samples of the polymer in the liquid and determining the change in weight of the polymer at equilibrium. Considerable information on retension and chemical and physical effects of water on polymers is available.

3.4 Tests for Diffusion and Permeability

Polymers are permeable to liquids having similar solubility parameters. Several different cells have been designed to measure the permeability of membranes to gases and vapors.

Permeability to vapors is determined by measuring the amount of vapor passage through a film of unit thickness and area in a specific period of time. The permeability of a polymer to a specific liquid may be determined by measuring, over a period of time, the change in weight of a freshly filled bottle molded from the polymer (ASTM-D2684). The permeability factor P is equal to the product of the weight change per day R and the bottle thickness T divided by the bottle surface area A.

The water vapor transmission WVT of a polymer sheet is determined by fastening the sheet over a dish containing water and measuring the loss in weight of water in a specified time (ASTM-E96).

Water absorption is measured by immersing water-insoluble polymer samples in water and determining the increase in weight after a 24-h period (ASTM-D570).

3.5   Tests for Mechanical Properties

There are a variety of methods which are useful in predicting mechanical performance properties. Five of the more commonly employed techniques are shown in Figure 3.1.

The tensile strength test [Figure 3.1 (1)] (ASTM-638) employs samples of a specified shape, typically a dogbone, as depicted in Figure 3.2. The sample is clamped at one end and pulled at a constant rate of elongation until the center of the specimen fails.

The length of the center section is called the initial gauge length L_0. The force F is measured at the fixed end as a function of elongation. Typically one plots stress S as a function of strain γ; see Eqs. (3.1) and (3.2), where A_0 is the original, undeformed cross-sectional area of the gauge region and ΔL is the change in sample length as a result of the applied force.

$$S = F/A_0 \qquad (3.1)$$

$$\gamma = \Delta L/L_0 \qquad (3.2)$$

A true or real stress is calculated as the ratio of measured force to instantaneous change A, i.e., dA, at a given elongation and is a more accurate measure of sample performance.* The real stress S_r is then

$$S_r = F/\mathrm{d}A = (F/A_0)(L/L_0) \qquad (3.3)$$

where L is the final length.

---

* The actual area may be determined from the original cross-sectional area, A_0, and measured elongation through an assumption of constant sample volume.

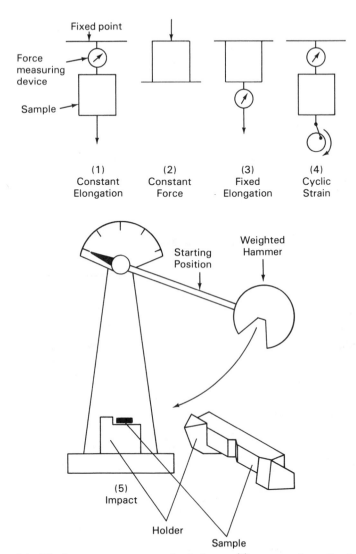

Figure 3.1 The five most common mechanical tests: (1) constant elongation for tensile strength determinations, (2) constant force for creep determinations, (3) fixed elongation for stress relaxation determinations, (4) cyclic strain for dynamic mechanical determinations, and (5) impact for impact determinations. (After J. Fried, *Plastics Engineering*, July 1982, with permission.)

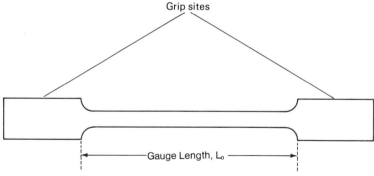

Figure 3.2 Typical ASTM tensile-test bar.

For elastomers exhibiting high extensions, real strain is defined as

$$\gamma_r = \ln(L/L_0) \tag{3.4}$$

The assumption that volume remains constant during deformation is not entirely correct; the change in volume is related to the extent of cross-linking, etc. For most polymers this change in volume ΔV is related to Poisson's ratio P as follows:

$$\Delta V = (1 - 2P)\gamma V_0 \tag{3.5}$$

The Poisson ratio is typically between 0.3 and 0.4, approaching 0.5 for incompressible materials.

Figure 3.3 shows representative stress–strain curves for a variety of polymeric materials. At normal use temperatures, such as room temperature, rigid polymers such as polystyrene (PS) exhibit a rapid increase in stress with increasing strain until sample failure. This behavior is typical of brittle polymers with weak interchain secondary bonding. As shown in the top curve in Figure 3.3, the initial stress–strain relation in such polymers is approximately linear and can be described in terms of Hooke's law, i.e., $S = E\epsilon$, where E is Young's modulus, typically defined as the slope of the stress–strain plot. At higher stresses, the plot becomes nonlinear. The point at which this occurs is called the *proportional limit*.

Many engineering thermoplastics such as nylon, high-impact PS, polyesters, and toughened plastics exhibit responses similar to those shown in the next two curves in Figure 3.3, designated *ductile*. Here the stress achieves a maximum called a *yield stress* at a specific strain. As strain increases beyond

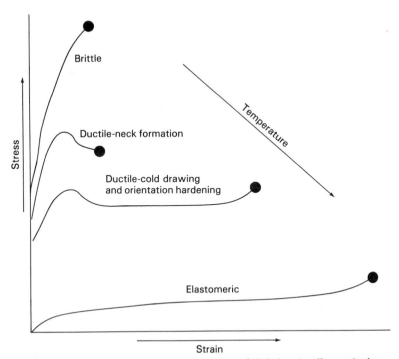

Figure 3.3 Stress–strain plots representative of brittle, ductile, and elastomeric polymeric materials. Failure is denoted by ●. (After J. Fried, *Plastics Engineering*, July 1982, with permission.)

this point, stress initially decreases (typically between 5 and 50% strain). The minimum stress at this point is called the *drawing stress*, and at this point the specimen may fail (second curve from the top) or undergo orientation hardening (third curve from the top) prior to specimen failure. Molecularly, polymer chains are stretched in the direction of load during orientation hardening. The initial portion of the stress–strain plot of the ductile materials, like that of the brittle materials, is linear, Hookean. The initial slope is generally less than that of the brittle materials, indicating a lower modulus for the ductile materials. However, the energy required to deform ductile materials to failure is typically higher than that required to deform brittle specimens, as indicated by a comparison of the areas under the stress–strain curves. This means that ductile polymers can absorb more energy upon impact without failure.

Elastomer behavior is depicted by the bottom curve in Figure 3.3. Here the modulus (ratio of stress to strain, as of strength to elongation; measure of polymer stiffness) is low, but elongations to several hundred percent are possible before failure.

Typical amorphous polymers can exhibit each of these types of stress–strain behavior when the temperature is changed from below to above the T_g of the polymer.

The dimensional stability of a polymeric material is generally determined by employing two types of transient measurements—creep and stress relaxation. Creep measurements are taken as the sample experiences application of a constant load [Figure 3.1 (2)]. Creep measurements can be made while employing flexure, compression, shear, tension, or torsion deformation. All forms are important in describing a material's performance with regard to sustaining loads for extended times. The usual property of interest is the ratio of strain to stress called the tensile compliance C. For creep measurements, stress is held constant with strain dependent on the time t the load is applied, giving C a time-dependence.

Although a number of assemblies have been developed for measuring creep, a simple, yet satisfactory, first approximation is obtained by clamping one end of the sample (of specified dimensions) to a stationary point and attaching the other end to a platform to which weight is rapidly added. Elongation is then measured and recorded as a function of time.

For an elastomer, C typically increases as temperature increases. This means the sample is softening as temperature increases, since C is proportional to the reciprocal of modulus, or stiffness.

Some applications require the material to remain under constant stress for years, yet it is often not reasonable to conduct such extended time measurements. One approach which circumvents this employs time–temperature superposition. Measurements are obtained over a shorter time span at differing temperatures. A master curve of C as a function of a reduced time t/a where a is a shift factor, is generated, and this allows the results to be extended to longer times. The shift factor is obtained by employing the Williams, Landel, and Ferry (WLF) relationship

$$\text{Log } a = \frac{-C_1(T - T_0)}{C_2 + T - T_0} \tag{3.6}$$

where C_1 and C_2 are constants and T_0 is the reference temperature. If T is the T_g, then C_1 and C_2 typically approach the values of 17.4 and 51.6, respectively.

A representative plot appears as Figure 3.4. At short times this polymer is somewhat elastic and partly viscous, at longer times it behaves as a rubber, and at the longest times it has the properties of a viscous liquid.

Stress relaxation measurements [Figure 3.1 (3)] are obtained by employing similar equipment. Here the force (weights, etc.) is rapidly applied

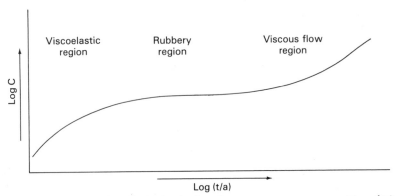

Figure 3.4   Creep plot for T_0 obtained using time–temperature superposition. (After J. Fried, *Plastics Engineering*, July 1982, with permission.)

and the stress measured as a function of time. Although the strain is a constant, the stress is time-dependent, as in the case of the tensile compliance C. Data handling, similar to that described for creep, and employed for tensile measurements, can be carried out giving a master curve such as that depicted in Figure 3.5. Here the time-related stress relaxation modulus $R(t)$, is defined as follows:

$$R(t) = S(t)/\gamma° \tag{3.7}$$

Dynamic testing [Figure 3.1 (4)] typically measures the stress response

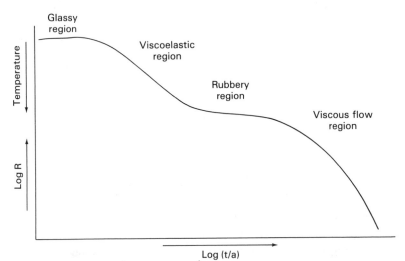

Figure 3.5   Stress relaxation plot for T_0 obtained using time-temperature superposition. (After J. Fried, *Plastics Engineering*, July 1982, with permission.)

of a specimen subjected to a strain that is varied with time. When strain is a sinusoidal function of time, strain can be described in terms of the frequency of oscillation f (radians per second) as

$$\gamma(f) = \gamma_0 \sin ft \qquad (3.8)$$

where $\gamma(f)$ is the strain.

The resulting stress can likewise be described by

$$S(f) = S_0 \sin(ft + p) \qquad (3.9)$$

where S_0 is the amplitude of the stress response at zero frequency of oscillation and p is the phase angle between the stress and the strain.

For Hookean materials the strain and the stress are in phase, and $p = 0$, whereas for viscous liquids the stress response trails the strain.

With one of the simplest assemblies for dynamic testing, the specimen is clamped at one end, which remains fixed [Figure 3.1 (4)]. The other end is connected to an inertia disk. The disk is set in motion, and the polymer dampens the oscillation; the dampening is characteristic of the polymer. Disk oscillations are typically on the order of 10^{-2} to 50 Hz.

Cyclic fatigue measurements require the specimen to be subjected to cyclic stress or strain of a higher amplitude than that employed for the simple dynamic test just described. The deformation must be of sufficient intensity to bring about specimen failure after a certain number of cycles, N. The value of stress leading to failure for a given N is typically 20% to 40% of the static tensile strength.

The fatigue life is the number of cycles a specimen can remain at a specified strain or stress before specimen failure occurs. As stress is decreased, there is a point beyond which failure does not occur regardless of the number of cycles the specimen experiences. This stress value is called the *endurance limit* or endurance strength.

Impact tests [Figure 3.1 (5)] measure the energy required for failure when a standard specimen receives a rapid stress loading. The impact strength of a polymer can be measured employing a number of techniques including the Izod and the Charpy tests. For both the Izod and Charpy tests a weight is released, causing the specimens to be struck [Figure 3.1 (5)]. The energy-to-break values are determined from the loss in the kinetic energy of the weight. Brittle polymers such as PS typically have low impact strengths, whereas many engineering thermoplastics such as polyethylene terephthalate (PET) and polycarbonate (PC) have high impact values.

The hardness of a polymer may be measured by the Rockwell tests, in which steel balls of specified diameter are used to indent the polymer specimen (ASTM-D785). The indentation hardness of rubbery polymers is measured by an indenter called a durometer (a rounded surface of specified weight and dimensions) (ASTM-D2240) which acts as the indenter onto test specimens.

Resistance to abrasion is measured by determining the weight loss at the abraded surface. Scratch resistance may be measured by using materials of known hardness to scratch the polymer surface. Pencils with known degrees of hardness may be used for this test.

3.6   Tests for Optical Properties

The index of refraction of transparent plastics may be determined by placing a drop of a specified liquid on the surface of the polymer before measuring the index with a refractometer. An optical microscope is used to measure the index of refraction in an alternative method (ASTM-D542).

A Hardy-type spectrophotometer may be used for determining luminous reflectance, transmittance, and color of polymers (ASTM-791). The transmittance of plastic films is measured by ASTM-D1746.

3.7   Flammability Tests

Small-scale horizontal flame tests have been used to estimate the flammability of solid (ASTM-D635) and foamed polymers (ASTM-D1992), but these tests are useful for comparative purposes only. Large-scale tunnel tests (ASTM-E84) and corner wall tests are more significant, but they are also more expensive than laboratory tests.

One of the most useful laboratory flammability tests is the oxygen index (OI) test (ASTM-D2043). In this test, the specimen is burned as a candle in controlled mixtures of oxygen and nitrogen. The minimum oxygen concentration which produces downward flame propagation is considered the OI of ignitability for the polymer.

3.8   Tests for Electric Properties

The dielectric constant (permittivity) and loss index are determined by ASTM-D150. Permittivity is the ratio of the capacitance of the polymer to that of air.

The dielectric breakdown voltage or dielectric strength of polymers may be determined by ASTM-D149. The dielectric breakdown voltage is the maximum applied voltage that a polymer can withstand for 1 min divided by the thickness of the polymer.

The electric resistance or conductance of a polymer is determined by measuring the dc voltage drop under specified conditions (ASTM-D257).

The power factor (dissipation factor) is the energy required to rotate the dipoles of a polymer in an applied electrostatic field of increasing frequency (ASTM-D150). The loss factor is equal to the product of the power factor and the dielectric constant of the polymer.

The arc resistance (ASTM-D495) is a measure of the ability of the polymer to resist the action of a high-voltage arc on the surface of the polymer. The results of this test may be used to classify materials in the laboratory.

3.9 Tests for Chemical Resistance

The classic test for chemical resistance (ASTM-D543) measures the percentage weight change (PWC) of test samples after immersion in many different liquid systems. Tests for chemical resistance have been extended to include changes in mechanical properties of the polymer test sample after immersion. Although there is no standard test of changes in mechanical properties of the samples, changes in the following have been investigated: hardness, tensile strength, stress relaxation, stress rupture, impact strength, compressive strength, flexural strength, and flexural modulus. Since chemical attack involves changes in chemical structure, it can be readily observed by many instrumental methods that measure chemical structure.

3.10 Weatherability

Outdoor exposure of polymer samples, mounted at a 45° angle and facing south, has been used to measure the resistance of polymers to outdoor weathering (ASTM-D1345). Since these tests are expensive and time-consuming, tests such as ASTM-G23 have been developed in an attempt to gain "accelerated test" results.

There are several accelerated tests which differ in the selection of light source and cyclic exposure to varying degrees of humidity. Some accelerated tests include salt spray, heat, cold, and other weather factors.

References on the testing of polymers are listed in Sec. 3.11.

3.11 References

1. *ASTM 1980 Annual Book of Standards*, Parts 35–38. Pipe and Building Products, General Test Methods, Part 36, Film, reinforced and cellular plastics, high modular fiber and composites Part 37, 38, published by American Society for Testing Materials, Philadelphia, Pennsylvania.

2. J. Chiu, *Polymer Characterizations by Thermal Methods of Analysis*, Dekker, New York (1974).

3. G. Gee, Thermodynamics of rubber solutions and gels, in *Advances in Colloid Science*, Vol. 2, Interscience, New York (1946).

4. S. P. Rowland, *Water in Polymers*, ACS Symposium Series 127, American Chemical Society, Washington, D.C. (1980).

5. S. B. Tuwinner, L. P. Miller, and W. E. Brown, *Diffusion and Membrane Technology*, ACS Monograph Series 156, Reinhold, New York (1962).

6. R. B. Seymour and C. E. Carraher, Chapter 5 in *Polymer Chemistry: An Introduction*, Dekker, New York (1981).

7. R. B. Seymour, *Plastics vs. Corrosives*, Chapter 7, Wiley-Interscience, New York (1982).

8. W. E. Driver, *Plastic Chemistry and Technology*, Chapter 12, Van Nostrand-Reinhold, New York (1979).

9. K. C. Frisch and A. V. Patsis, *Electrical Properties of Polymers*, Technomic, Westport, Conn. (1972).

10. L. H. Lee, *Advances in Polymer Friction and Wear*, Plenum Press, New York (1974).

11. R. B. Seymour, *Modern Plastics Technology*, Reston Publishing, Reston, Va. (1975).

12. O. H. Varga, *Stress-Strain Behavior of Elastic Materials*, Wiley-Interscience, New York (1966).

4 | Optical Properties of Polymers

4.1 General

Since polymers are often used as clear plastics or coatings and have many applications in which transparency is an important property, a knowledge of the optical properties of specific polymers is essential. The radiation scale, of course, includes microwave, infrared, ultraviolet, and visible regions, but the emphasis in this chapter is on the latter.

The molecular behavior in these regions varies: the energy associated with the microwave wavelength region corresponds to the energy necessary to bring about rotational motions; energy associated with the infrared region corresponds to energy necessary to effect vibrational motions; and energy associated with the ultraviolet spectral region corresponds to energy necessary to dislocate electrons. Responses to radiation in all the regions may be employed for structural and property analyses. The relations between material electric responses and the type of radiation employed are discussed in Chapter 6. Material scientists are concerned primarily with responses in the visible region which corresponds to the upper (more energetic) infrared and lower ultraviolet energy regions.

The amount of radiation used is also important. A Nernst glower is used in ordinary infrared spectroscopy. This light source emits a relatively low amount of radiation, and no destruction of the analyzed material occurs. However, Raman infrared spectroscopy employs a radiation source of much greater energy. This radiation is sufficiently energetic to cause bond disruption and some destruction of the analyzed material.

It is important to recognize the difference between refraction (associated with properties such as refractive index) and reflection (associated with properties such as haze). This difference is illustrated in Figure 4.1.

4.2   Refractive Index

Optical properties are related to both the degree of crystallinity and the actual polymer structure. Most polymers are transparent and colorless, but some, such as phenolic resins and polacetylenes, are colored, translucent, or opaque. Polymers that are transparent to visible light may be colored by the addition of colorants, and some become opaque as the result of the presence of additives, such as fillers, stabilizers, flame retardants, moisture, and gases.

Many of the optical properties of a polymer are related to the refractive index n, which is a measure of the ability of the polymer to refract or bend light as it passes through the polymer. The index n is equal to the ratio of the sine of the angles of incidence, i, and refraction, r, of light passing through the polymer:

$$n = \frac{\sin i}{\sin r} \qquad (4.1)$$

The magnitude of n is related to the density of the substance and varies from 1.000 and 1.333 for air and water, respectively, to about 1.5 for many polymers and 2.5 for the white pigment titanium dioxide. The value of n is high for crystals and is dependent on the wavelength of the incident light and the temperature; it is usually reported for the wavelength of the trans-

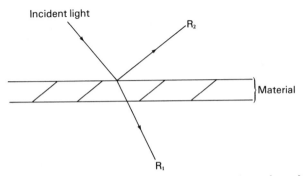

Figure 4.1.   Refraction and reflection of incident light at the surface of a solid. The path of refracted light is indicated by R_1 and that of reflected light by R_2.

parent sodium D line at 298 K. The n_D values of polymethyl methacrylate (PMMA) and polystyrene (PS) are 1.490 and 1.592, respectively.

The velocity of light passing through a polymer is affected by the polarity of the bonds in the molecule. Polarizability P is related to the molecular weight per unit volume, M, and density ρ as follows (the Lorenz-Lorentz relationship):

$$P = \left(\frac{n^2 - 1}{n^2 + 2}\right)\frac{M}{\rho} \qquad (4.2)$$

The polarizability P of a polymer is related to the number of molecules present per unit volume, and the polarizability of each molecule is related to the number and mobility of the electrons present in the molecule. The P value of carbon is much greater than that of hydrogen, and the value of the latter is usually ignored in calculating the polarizability of organic polymers.

Representative n values of selected polymers are shown in Table 4.1. The n values of most polymers are between 1.40 and 1.70, and these values

Table 4.1 Refractive Indices of Selected Polymers

Polymer	Refractive index (25 °C)[a]
Polytetrafluoroethylene (PTFE)	1.35
Silicone	1.43
Polymethylpentene (TPX)	1.46
Cellulose acetate butyrate	1.47
Acetal polymer	1.48
Polyvinyl butyral	1.48
Polypropylene (PP)	1.49
Polymethyl methacrylate (PMMA)	1.50
Polyallyl diglycol carbonate	1.50
Polybutylene	1.50
Polyethylene (LDPE)	1.51
Polyvinyl chloride (PVC)	1.53
Unsaturated polyester	1.54
Polyethylene (HDPE)	1.54
Polyurethane (PU)	1.55
Urea resin	1.55
Epoxy resin	1.57
Polycarbonate (PC)	1.58
Polystyrene (PS)	1.59
Phenolic resin	1.60
Polyvinylidene chloride (PVDC)	1.62
Polyaryl sulfone	1.67

[a]Wavelength = 400 nm.

usually increase as the number of polarizing groups present in the molecule increases. If a material is structurally isotropic, as in the case of an unstressed amorphous polymer, then it is optically isotropic, and a single n value (and other related optical properties) can be assigned. In contrast, the optical properties of an anisotropic material, such as a crystalline polymer, vary along different principal axes, and the material is referred to as having *double refractance* or *birefringence*.

An isotropic polymeric material may exhibit birefringence as a result of the application of stress which causes preferential realignment of the molecules along the stress crease, and this causes an anisotropic polarization of incoming electric, thermal, or visible radiation. The resulting birefringence can be used to measure the effects of deformation. In photoelastic stress analysis, the material is viewed by transmitting light between crossed polaroids.

Birefringence induced by applied stress is caused by the two components of the refracted light traveling at different velocities. This generates interference which is characteristic of the material. The change in refractive index, Δn, produced by a stress S is often related by a factor C called the stress-optical coefficient as follows:

$$\Delta n = CS \qquad (4.3)$$

The n value also varies with the wavelength of the incident radiation. Thus the n value of PMMA is 1.505 at a wavelength of 400 nm and 1.485 at 700 nm.

4.3 Optical Clarity

As shown by the Lambert-Beer law,

$$\mathrm{Log}\,\frac{I}{I_0} = -AL \qquad (4.4)$$

optical clarity, or the fraction of illumination transmitted through a polymer, I/I_0, is dependent on the path length of the light, L, and the absorptivity of the polymer at that wavelength of light, A.

Clarity is noted when the light passes through a homogeneous sample, such as a crystalline, ordered polymer or a completely pure amorphous phase. Interference occurs when the light beam passes through a heterogeneous sample in which the polarizability of the individual units varies slightly, such as a polymer containing both crystalline and amorphous regions.

Thus commercial atactic PS is transparent, but isotactic PS is translucent because of some amorphous areas present in the highly crystalline product. Also, high-impact PS is translucent because of the presence of two phases.

Refraction, R_0, which is related to the "hiding power" of a pigment, is equal to the square of the differences in the n values of the pigment and the polymer, as shown by Fresnel's law where n_1 and n_2 are the refractive indices of the pigment and the polymer.

$$R_0 = \left(\frac{n_1 - n_2}{n_1 + n_2}\right)^2 \qquad (4.5)$$

Pigments such as titanium dioxide ($n = 2.5$) with n values which are higher than that of the polymer are effective opacifiers. The opacity is also related to the size of the pigment particles.

Translucent plastics may be produced by the addition of fillers such as alumina trihydrate (ATH) which have n values similar to that of the polymer. Such fillers also serve as flatting agents in paints.

The opacity of plastic foams, and polymers with scratched surfaces, is also governed by Fresnel's law. The n value of the gas which occupies the scratch indentation is much lower than that of the polymer. Light may be directed through rods of transparent polymers, such as PMMA. This effect may be enhanced when the rod or filament is coated with a polymer with a different refractive index, such as polytetrafluoroethylene (PTFE). Optical fibers utilize this principle.

The density and n value of a polymer crystal are greater than those of an amorphous polymer. Many polymers are opaque because of the presence of ordered clusters of crystals called spherulites which have different n values. PTFE, which is highly crystalline, is opaque; but amorphous polycarbonate (PC), PMMA, and PS are noncrystalline and clear.

The clarity of a translucent crystalline polymer, such as polypropylene (PP), may be improved by biaxial orientation. Monoaxial orientation of a crystalline film produces an anisotropic birefringent film.

Alignment of polymer chains during tensile strength testing also results in the product becoming birefringent. Birefringence increases if annealing occurs under stress but decreases if the substance is annealed under non-stressed conditions.

The clarity of a translucent type of film, such as a low-density polyethylene (LDPE) film, may be improved by quenching a molten film. This improvement, which is the result of the formation of smaller than usual spherulites, is enhanced by the presence of nucleating agents, such as benzoic acid.

The opacity of a crystalline polymer may also be reduced by copolymerization. Even so, polymethylpentene (TPX) is transparent because the n values of the amorphous and the crystalline phase are quite similar. Transparency may also be enhanced by copolymerization of the methylpentene with a small amount of another polymerizable olefin with a similar n.

4.4   Absorption and Reflectance

Colorless materials range from being almost totally transparent to being opaque. The opacity is related to the light-scattering process occurring within the material.

The incident radiation passes through nonabsorbing, isotropic and optically homogeneous samples with essentially little loss of radiation intensity. Actually, all materials scatter some light. The angular distribution of the scattered light is complex because of the scattering due to micromolecular differences in n values. Yet related parameters may be employed as measures of this lack of homogeneity.

Transparency is defined as the state permitting perception of objects through a sample. Transmission is the light transmitted. In more specific terms transparency, T, is the amount of undeviated light, i.e., $T =$ original intensity minus all light absorbed, scattered, or lost through any other means. The ratio of reflected light to the incidental light is called the reflectance coefficient, RC, and the ratio of the scattered light to the incident light is called the absorption coefficient, AC. The attenuation coefficient is the name associated with the amount of light lost and is typically the sum of RC and AC. The transmission factor, TF, is the ratio of the amount of transmitted light to the amount of incident light. The TF for weakly scattering colorless materials decreases exponentially with the sample thickness l and the scattering coefficient S.

$$\ln TF = \text{sl} \qquad (4.6)$$

This light scattering reduces the contrast between light, dark, and other colored parts of objects viewed through the material and produces a milkiness or haze in the transmitted image. Haze is a measure of the amount of light deviating from the direction of transmittancy of the light by at least 2.5°.

Heterogeneity of n values is related to a number of factors, including end groups, differences in density between amorphous and crystalline regions, anisotropic behavior of crystalline portions, incorporation of additives, and the presence of voids.

The visual appearance and optical properties of a material depend on its color and additives, etc., as well as on the nature of its surface. *Gloss* is a term employed to describe the surface character of a material which is responsible for luster or shine, i.e., surface reflection of a polymeric material.

The reflection R'_0 is related to the magnitude of the difference between the n value of the polymer, n_1, and that of air, n_2, as shown by Fresnel's law, Eq. (4.5). A perfect mirror-like surface reflects all incident light, and this represents one extreme. At the other extreme is a highly scattering surface which reflects light equally in all directions at all angles of incidence. The direct reflection factor is the ratio of the light reflected at the specular angle to the incident light for angles of incidence from 0 to 90°.

Total reflectance is observed at angle θ_B (Brewster's angle), which is related to the ratio of the n values of the polymer, n_1, and air, n_2.

$$\text{Tan } \theta_B = \frac{n_1}{n_2} \qquad (4.7)$$

A number of different types of optical behavior with respect to incidence and specular light radiation are shown in Figure 4.2.

Charring of polymers, such as PS, is accompanied by a darkening of the residue through the formation of highly aromatic intermediate networks.

Black polymers, such as polyacetylene, absorb all visible light, but opaque polymers scatter the incident light. As shown by Lord Raleigh, the turbidity τ is related to the scattered light per unit volume, i_0, integrated over all angles.

$$\tau = \int_0^\pi 2\pi i_0 \sin \theta \; d\theta \qquad (4.8)$$

The amount of excess light scattered by dilute polymer solutions compared with that scattered by the pure solvent is used routinely to determine the weight average molecular weight of polymers, $\overline{M}_w$.

Absorption of light or loss of intensity I of light passing through a path of distance l may be calculated from the Lambert relationship

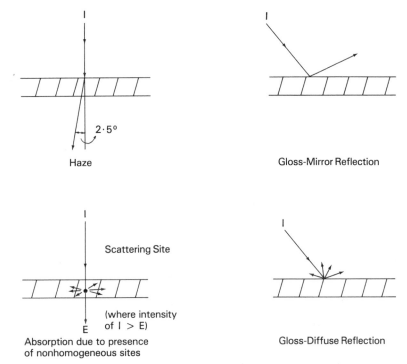

Figure 4.2. Types of optical behavior of polymeric materials.

$$I = I_0 e^{-4\pi Nkl/\lambda} \tag{4.9}$$

where k is the absorption index, λ is the wavelength of the light, I_0 is the original intensity, and N is Avogadro's number.

Most polymers absorb electromagnetic radiation at characteristic wavelengths in the infrared region. Hence infrared spectroscopy is used for the

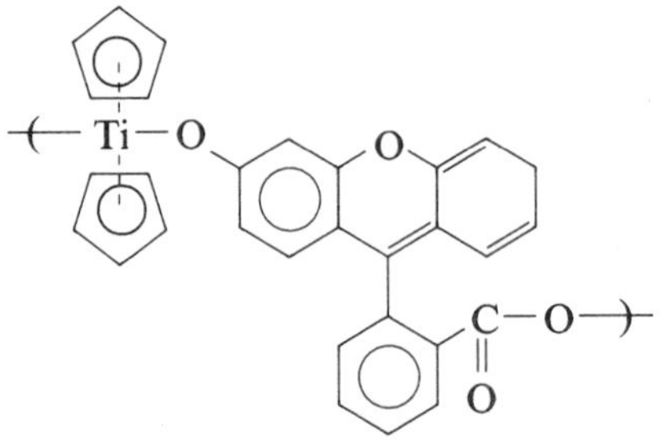

Figure 4.3. Structure of a titanium polydye.

qualitative and quantitative analysis of polymers. Infrared spectroscopy may also be used to determine the extent of crystallinity in a polymer.

As described in Chapter 6, Electric Properties of Polymers, there is a general relationship between the delocalization of electrons throughout a polymer chain or network and color so that the incidence of and darkness of color increases as electron delocalization increases. Thus polyethylene is colorless while polyacetylene is black.

Color sites can be introduced either through inclusion of monomers possessing color sites as in the case of titanocene polyfluorocene dyes, shown in Figure 4.3, or by the generation of color sites through polymerization, as in the case of polyphenylenes.

4.5 References

N. Abazajian and R. B. Seymour, *Handbook of Polymer Chemistry*, McGraw-Hill, New York (1984).

L. J. Bellamy, *Advances in Infrared Group Frequencies*, Methuen, London (1968).

G. I. Kinney, *Engineering Properties and Applications of Plastics*, Wiley, New York (1957).

M. L. Miller, *The Structure of Polymers*, Reinhold, New York (1966).

P. D. Ritchie, ed., *Physics of Polymers*, Iliffe, London (1965).

H. A. Szymanski, *Infrared Handbook*, Plenum Press, New York (1964).

P. W. Van Krevelin, *Properties of Polymers*, Chapter 12, Elsevier, New York (1972).

5 | Mechanical Properties of Polymers

5.1 Introduction

When a polymer is used as a structural material, it is important that it be capable of withstanding applied stresses and resultant strains over its useful service life. Polymers are viscoelastic materials, having the properties of solids and viscous liquids. These properties are time- and temperature-dependent.

Thus the effects of the rate of application of stress and the ambient temperature must be recognized when polymers are used as structural materials, and definite rates and temperatures must be specified for tests, such as those for tensile and flexural strengths cited in Chapter 3. A knowledge of the structure of polymers is essential for the understanding of these effects, which differ from the effects of stress and temperature on all other materials of construction.

Numerous factors affect various mechanical properties of polymers, including molecular weight, processing, extent and distribution of crystallinity, composition of polymer, and use temperature.

Chapter 3 describes basic mechanical testing procedures. This chapter describes various structural-related factors which affect polymeric mechanical properties.

5.2 Molecular Weight

Many physical and mechanical properties of amorphous polymers improve rapidly as the molecular weight increases up to the threshold molecular weight (Sec. 2.2). However, the change levels off after a moderately high molecular weight is reached (Figure 5.1).

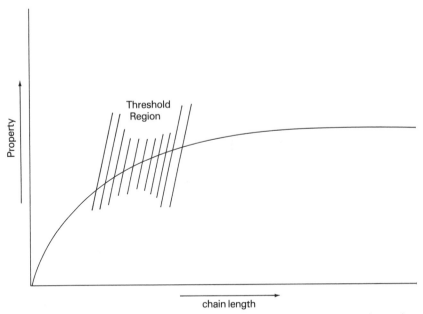

Figure 5.1 Relation between chain length and select polymeric properties such as T_g, yield stress, and modulus.

In contrast, the viscosity increases exponentially as the molecular weight increases above the threshold molecular weight. Since more energy is required to process these high-molecular-weight polymers, an optimum or commercial range is often selected for commercial general purpose polymers.

Although the average molecular weight of synthetic elastomers (SRs) is within the commercial range, the molecular weight of natural rubber (NR) is over 1 million, which is above this arbitrary range. Hence NR is masticated in a mill or an intensive mixer in order to cleave the long chains and reduce the average molecular weight. Chain transfer agents called peptizers may be added to the mix to prevent recombination of the macroradicals produced in this mechanical degradation process.

Most thermoplastic materials have high molecular weights, beyond the threshold region, so moderate changes in molecular weight do not appreciably influence mechanical properties such as yield stress or modulus. Even so, rupture properties such as ultimate elongation, impact strength, and ultimate strength are affected by molecular weight increases since longer chains are more likely to act as connectors between crystalline areas, allowing greater toughness.

Lower-molecular-weight polymers with molecular weights above that required for entanglement are more flexible than higher-molecular-weight polymers because of the greater number of chain ends per unit volume in the lower-molecular-weight polymers. The chain ends reduce the packing efficiency of the lower-molecular-weight polymers.

In spite of processing difficulties, some higher-molecular-weight polymers, such as ultrahigh-molecular-weight polyethylene (UHMWPE), are used because of their superior toughness.

The modulus, or stiffness, of a polymer is also a function of intermolecular forces, stiffening groups, and additives, such as reinforcing fillers. Most fibers have intermolecular hydrogen bonds, and many engineering plastics have stiffening groups, such as phenylene groups, in the polymer backbone.

The modulus of polymers also increases as the cross-link density of network polymers is increased. Crystalline and filled polymers resemble polymers with low cross-link density at temperatures below the melting temperature T_m.

Although some polymers may be satisfactory when used under the stress of static loads, they may fail when subjected to impact. The impact resistance, or resistance to brittle fracture, is a function of the molecular weight of a polymer. Thus UHMWPE is much more resistant to impact failure than general purpose high-density polyethylene (HDPE). The impact resistance of brittle polymers is also increased by the addition of plasticizers. Thus polyvinyl chloride (PVC), plasticized by relatively large amounts of dioctyl phthalate, is much less brittle than unplasticized rigid PVC.

5.3 Crystallinity

Selected factors affecting crystallinity regarding T_g and T_m are described in Chapter 2. Here we discuss the influence of crystallinity on the mechanical properties of polymers. For thermoplastics the relation between the degree of crystallinity and the physical nature is shown in Table 5.1. The general lack of difference in physical nature shown by largely crystalline polymers at

Table 5.1   Effect of Crystallinity on Physical Appearance

	Degree of crystallinity		
	Low (< 10%) amorphous	Intermediate (20–70%)	High (> 70%) crystalline
Above T_g	Rubbery	Tough, leathery	Hard, stiff
Below T_g	Brittle, glassy	Tough, horny	Hard, stiff

temperatures below and above the T_g is because they lack a significant portion with a T_g.

Table 5.2 lists polymers and their tendency toward crystallinity. Yield stress and strength, and hardness increase with an increase in crystallinity as does elastic modulus and stiffness. Physical factors that increase crystallinity, such as slower cooling and annealing, also tend to increase the stiffness, hardness, and modulus of a polymeric material. Thus polymers with at least some degree of crystallinity are denser, stiffer, and stronger than amorphous polymers. However, the amorphous region contributes to the toughness and flexibility of polymers.

Polymers with little structural symmetry and with bulky pendant groups, such as atactic polystyrene (PS), are usually amorphous. Amorphous polymers have no long-range order, and their x-ray diffraction patterns are diffuse halos rather than the sharp peaks which are characteristic of crystalline polymers.

Linear polymers prepared by step reaction polymerization, such as nylon 66, and linear, ordered polymers prepared by the chain polymerization of symmetrical vinylidene monomers, such as polyvinylidene chloride (PVDC), can usually be crystallized because of symmetry and secondary-bonding. Isotactic polymers, such as isotactic polypropylene (PP), usually crystallize as helices.

Although atactic polymers with bulky pendant groups are usually amorphous, atactic polymers with small pendant groups, such as polyvinyl alcohol (PVA), may be made to be crystalline. Some polymers, such as PVC, may have long sequences of syndiotactic sequences in addition to atactic sequences, and hence are somewhat crystalline.

Because of the geometric regularity present, gutta-percha (*trans*-polyisoprene) and stretched NR from *Hevea brasiliensis* (*cis*-polyisoprene) are crystalline. Random copolymers are usually amorphous, but some may be crystalline if the comonomers, such as ethylene and tetrafluoroethylene, are similar in size. Block copolymers may have crystalline domains if either of the com-

Table 5.2 Tendency for Crystalline Formation

High tendency	Moderate	Low
HDPE	PVC	Atactic polymers
PP (stereoregular)	NR	ABS
PTFE	*it*-PS	Random copolymers
Cellulose	*st*-PS	PMMA
PEO		(Most) thermosets
POM		PC
Nylon		

onomers produces crystalline homopolymers. The degree of crystallinity in block copolymers is related to the size of the crystallizable domain.

Crystalline condensation polymers, such as nylon 66, are useful at temperatures below the T_m, whereas amorphous addition polymers like atactic PS are useful as structural materials only at temperatures below the glass transition temperature T_g. In contrast, elastomers, such as vulcanized *cis*-polyisoprene (NR), are useful at temperatures above the T_g and the T_m. This is a consequence of the higher cohesive, secondary-bonding energies typically present (per unit) in condensation polymers which permit aggregate action to occur even above the T_g. It is important to again point out that single polymer crystals are customarily assigned only a T_m while totally amorphous polymers are assigned only a T_g. Polymers that contain both crystalline and amorphous regions are assigned both T_g and T_m transitions.

Although most fibers are derived from condensation polymers which exhibit strong intermolecular hydrogen bonding, some symmetrical vinyl polymers, such as isotactic PP, also form good fibers, because of the tight, symmetrical packing which can occur, permitting stronger intermolecular interactions. Because of strong segmental motion, crystallizable polymers do not tend to crystallize readily at temperatures within 10 K of the T_m. Also the high viscosity of the polymer–melts at temperatures within 10 K of T_g hampers the crystallization process. Thus the optimum temperatures for crystallization from a polymer melt are between $T_m - 10$ K and $T_g + 30$ K.

Since at temperatures below the T_g the chains of an amorphous polymer are randomly distributed and immobile, the polymers are typically transparent. These glassy polymers behave like a spring and when subjected to stress, can store energy in a reversible process. However, when the polymers are at temperatures slightly above the T_g, i.e., in the leathery region, unless cross-links are present, stress produces an irreversible deformation.

It is important to note that although many scientists emphasize differences in chemical structure and associated secondary bonding as a means to divide polymers into categories such as elastomers, fibers, and plastics, many polymers can belong to more than one category. Thus PP, PVC, polyacrylonitrile (PAN), polyethylene terephthalate (PET), and many nylons can be employed as both fibers and plastics. Further, although elastomers should ideally have low amounts of hydrogen bonding and polarity, some materials such as copolymers of acrylonitrile and butadiene (NBR), and polyurethanes (PUs) are good elastomers, yet have high polarities.

Thus although the property–structure relations are real, generalizations must not be blindly applied since polymer requirements can be met by materials which may not exhibit all the general characteristics of a given category.

5.4 Temperature

Structural relations influencing T_g and T_m are discussed in Chapters 2 and 7. It is important to note that most linear polymers are hard brittle plastics at temperatures below their characteristic T_g, leathery and rubbery at temperatures just above the T_g, and viscous liquids at temperatures above the T_m. In general, polymers are classified as plastics, fibers, and elastomers in accordance with the room-temperature relations of their T_g values. Some polymers, including network, highly cross-linked, and highly crystalline polymers, are difficult to melt and often undergo solid phase thermal degradation before melting occurs.

In the rubbery region, which is just above (in terms of temperature) the leathery region, polymer chains have high mobility and may assume many different conformations, such as compact coils, by bond rotation and without much disentanglement. When these rubbery polymers are elongated rapidly, they snap back in a reversible process when the tension is removed. This elasticity can be preserved over long periods of time if occasional cross-links are present, as in vulcanized soft rubber, but the process is not reversible for linear polymers when the stress is applied over long periods of time.

Rubber elasticity, which is a unique characteristic of polymers, is due to the presence of long chains existing in a temperature range between the T_g and the T_m. The requirements for rubbery elasticity are (1) a network polymer with low cross-link density, (2) flexible segments which can rotate freely in the polymer chain, and (3) no volume or internal energy change during reversible deformation.

Since there is no change in internal energy when an ideal elastomer is stretched, the entire contribution to the retraction or restoring force is entropy. Unstretched elastomers are amorphous, but the random chains become more ordered when the elastomer is stretched. The modulus of an elastomer changes slightly as the temperature is reduced, but there is an abrupt change in modulus as the elastomer becomes a glassy polymer at the T_g.

Most thermoplastics, such as PVC, have relatively high moduli at their useful temperatures, which are below the T_g. The intermolecular forces of plastics are much stronger than those of elastomers and are usually weaker than the intermolecular hydrogen bonds characteristic of most fibers. Figure 5.2 describes the idealized behavior of a thermoplastic as a function of chain length and temperature.

Many thermoplastics retain some molecular mobility below the T_g because of a secondary relaxation related to side chains or localized unoriented

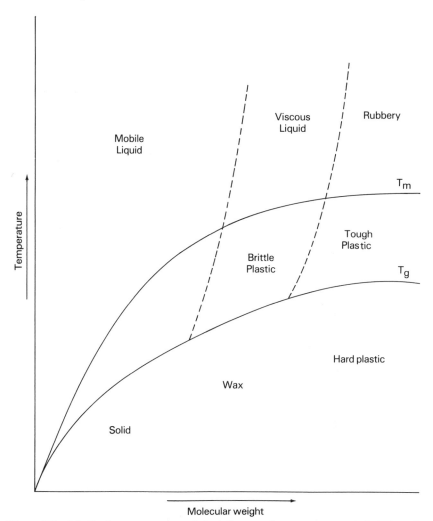

Figure 5.2   Idealized representation of the influence of temperature and chain length on a thermoplastic.

main chain segments. The terms *beta* and *gamma transition temperatures* are used to designate the temperatures at which these secondary relaxations cease.

The modulus of thermosets, such as phenolic plastics, is much greater than that of thermoplastics because of the high cross-link density present in thermosets. Since these network polymers are brittle, they are usually toughened by the addition of fibrous reinforcements.

It is customary to process thermosets as low-molecular-weight prepolymers and to cure or cross-link the prepolymers in a mold or on a surface. These cross-linked polymers do not melt or flow below their decomposition temperatures.

In fibers and drawn films, the polymer chains are aligned, so the polymers are anisotropic. Thus the moduli are many (to 30-fold) times greater in the longitudinal than in the trans direction. Although the high longitudinal strength of fibers is advantageous, high longitudinal strength is not always advantageous for uniaxially drawn film. Hence the anisotropy in the plane of the film is usually reduced by biaxial orientation. Biaxially oriented film, such as polypropylene, has much greater tear resistance in the longitudinal direction than uniaxially drawn film.

5.5. Polar Substituents

In general, substitution of polar atoms and polar groups for nonpolar or less polar moieties results in an increase in the T_g and such mechanical properties as yield stress and modulus. Thus condensation polymers such as nylons, polycarbonate (PC), and polyesters are typically higher-melting and exhibit higher T_g's, tensile strength and associated properties, but typically lower impact strengths and associated properties which require some flexibility (Table 5.3).

5.6. Additives

In general, the absorption of smaller molecules such as water acts to decrease the intermolecular forces between chains. Absorption is an attraction property in which likes attract. Thus nonpolar polymers such as PS, poly-

Table 5.3 Influence of Chain Structure on Select Mechanical Properties[a]

	T_g(°C)	T_m(°C)	Tensile strength (psi)	Tensile modulus(10^5) (psi)	Impact strength (ft · lb/in. of notch)
HDPE	−125	135	4 000	1.0	6
LDPE	−125	110	2 000	0.25	> 15
PP	−10	176	5 000	1.5	4
PS	100	240	7 000	4.5	0.3
PMMA	105	200	10 000	4.5	0.4
Nylon 66	57	265	10 000	3.5	1.0
PET	69	265			
PC			9 000	3.2	14

[a]Average values of ranges.

tetrafluoroethylene (PTFE), PE (polyethylene), and PP absorb nonpolar gases and liquids such as gasoline, hydrocarbon lubricating oils, and organic solvents such as chloroform yet are largely immune to water and ethanol. On the other hand, polar polymers such as PU, nylons, PC, and cellulosic products readily absorb water, the effect being to disrupt hydrogen bonding between chains. In general, properties requiring strong interchain bonding, such as tensile strength, flexural modulus, and yield strength, are decreased as water is added, whereas properties assisted by chain flexibility and "slickness," such as elongation at break and elongation at yield, are increased.

The addition of smaller molecules such as plasticizers and the entrapment of monomers have an effect similar to that of the absorption of water. In general, the modulus temperature transition (temperature at which polymer stiffness changes abruptly with the polymer being stiffer below the T_g) and the T_g decrease as the amount of additive is increased. Thus the modulus temperature transition occurs at about 100 °C for PVC itself, 70 °C for PVC containing 10% dioctyl phthalate, and 20 °C for PVC containing 30% dioctyl phthalate.

It should be noted that a plasticizer need not be an externally added smaller molecule but can be introduced as chain branches.

The toughness of brittle polymers, such as rigid PVC, may be improved by the addition of a semicompatible elastomer such as a polyethyl acrylate graft copolymer. Such composites are harder than polymers plasticized with liquid plasticizers.

Although cracks may propagate rapidly in homogeneous polymers, they are dispersed by the addition of elastomer particles forming plastic composites. The rubbery particles also provide an opportunity for chain mobility below the T_g of the polymer and this aids in relieving stress.

The durability of polymers may also be improved by the addition of reinforcing fillers and fibrous reinforcements (F) to the polymer matrix (M). The modulus of a composite is a function of the distribution (amount and orientation) of the fibers, f, the modulus of each component, G, and the partial volume of the fiber, C, as shown by the following equation:

$$G_{\text{composite}} = f[G_F C + G_M(1 - C)] \qquad (5.1a)$$

The value of f is 1 when the fibers are aligned and $1/3$ when the fibers are randomly oriented, i.e., isotropic composites. The value of f is decreased when the length of the fiber is less than the critical length (minimum chain length required for entanglement of the polymer chains).

When a load is placed on the composite, this load is transferred from the matrix or continuous phase to the fiber or discontinuous phase, providing there is a good interfacial bond between the two phases. Glass fiber is the most widely used reinforcement, but unusually strong composites are obtained when graphite, aramid (aromatic nylon), or boron fibers are used with polyesters, epoxies, or engineering thermoplastics.

5.7 Pressure

The application of pressure has a dramatic effect on the mechanical properties of polymeric materials. The precise nature of this effect varies with such factors as temperature, polymer, molecular weight and distribution, and magnitude of applied pressure. Each material must be evaluated separately under the specific conditions of use.

Even so several generalities are emerging. In contrast to metals, polymers generally show a large increase in modulus and yield stress with an increase in pressure, possibly because the pressure forces a realignment of chain segments and a closer association of these chains, resulting in stronger secondary bonding.

There is a variance in polymeric behavior with regard to fracture and the nature of yielding. Less brittle polymers, such as PVC, initially become more brittle as pressure is increased, with elongation falling to 60 ksi. Thereafter, elongation increases with increasing pressure. More brittle materials, such as PS and polyimide (PI), increase in ductility with an increase in pressure—a phenomenon similar to that observed for other brittle materials such as marble. With polymers such as PE and PP that show good cold drawing under atmospheric pressure, increasing pressure acts to restrict cold drawing, decreasing the nominal strain to fracture.

The above behavior may be envisioned as a consequence of two factors. First, pressure increases tend to reduce the free volume, and thus the chain mobility is lowered and the association between chains increased. Second, there appears to be a shift to higher temperatures of relaxation transitions as the pressure increases.

5.8. Physical Models

A number of attempts have been made to describe the behavior of polymers in terms of models. Following is a description of some of the simpler models employed in the description of the stress–strain behavior of polymeric materials. Although such models are insufficient to describe the behavior of even the simplest of polymeric materials, they are able to assist in the description of polymeric behavior on a molecular level.

Although such models leave much to be desired, they are able to indicate the type and relative importance of certain general types of behavior on a molecular scale.

When used in load-bearing applications, isotropic polymers may also fail because of low modulus. The moduli that must be considered in the design of functional polymers are Young's modulus E, shear modulus G, and bulk modulus K. Poisson's ratio P should also be considered.

The Poisson's ratio ($\Delta l / \Delta w$, where Δl is the change in length produced by a change of width, Δw) of an isotropic liquid is 0.5, and that of an elastic solid is about 0.2. The value of P for an elastomer, such as NR, is 0.5, and this value decreases as the elastomer is cured with increasing amounts of sulfur, i.e., as the crosslink density increases. Likewise, P for rigid PVC is about 0.3, and this value increases progressively as the plasticizer content is increased.

Poisson's ratio P is related to E and G as follows:

$$G = \frac{E}{2(1 + P)} \qquad (5.1b)$$

As shown by Eq. (5.2), K is proportional to E and related inversely to P.

$$K = \frac{E}{3(1 - 2P)} \qquad (5.2)$$

Young's modulus E is also related to K and G as shown in Eq. (5.3):

$$E = \frac{9KG}{3K + G} \qquad (5.3)$$

The effect of stress S on an ideal viscous liquid and on an ideal elastic solid may be predicted from Newton's

$$S = \eta \frac{d\gamma}{dt} \qquad (5.4)$$

and Hooke's

$$S = G\gamma \tag{5.5}$$

laws, respectively, where η = coefficient of viscosity, $\dfrac{d\gamma}{dt}$ = rate of strain, and γ = strain.

A spring with a modulus of G, and a dashpot containing a liquid with a viscosity of η, have been used as models for Hookean elastic solids and Newtonian liquids, respectively. In these models, the spring stores energy in a reversible process, and the dashpot dissipates energy as heat in an irreversible process. Figure 5.3 is a stress–strain curve for a typical elastomer; the straight

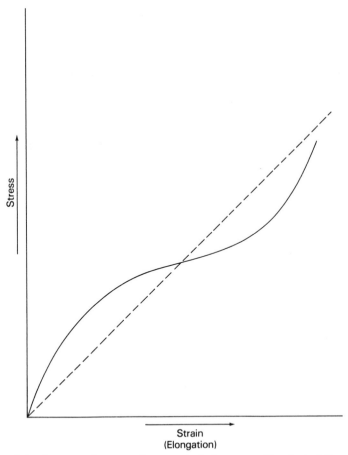

Figure 5.3 Stress–strain curve for a typical elastomer. The dashed line indicates Hookean behavior.

dashed line indicates Hookean behavior as described in Eq. (5.5). It is apparent that the elastomer behavior is not Hookean. A better model, but still unsatisfactory, is the Maxwell model, in which two simple elements, a Hookean spring and a Newtonian dashpot (Figure 5.4), are connected in series. Simple elastomer stretching involves the rapid, low-energy slippage of chains past one another as well as the breakage of the secondary bonds which act between the various chains: these two molecular features are roughly mimicked by the spring-dashpot series.

Over a century ago Maxwell used this combination of dashpot and spring to explain the mechanical behavior of pitch. The Maxwell model may be expressed mathematically as follows:

$$\frac{d\gamma}{dt} + \frac{G}{\eta} = 0 \tag{5.6}$$

$$S = S_0 e^{-t(G/\eta)} \tag{5.7}$$

where S_0 is equal to the initial stress before the stretched specimen is allowed to relax exponentially with time t. When $t = \eta/G$, the S is reduced to $1/e$ ($1/e = 1/2.7 = 0.37$) times the original value.

The relaxation time is equal to η/G.

As shown in Figure 5.4, when a Maxwell model is subjected to an applied stress, it first deforms instantaneously and then undergoes irreversible flow.

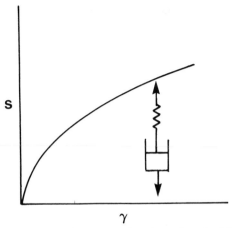

Figure 5.4   Stress–strain plot for stress relaxation using the Maxwell model.

In contrast, in a model proposed by Voigt and Kelvin (Figure 5.5), in which the spring and dashpot are in parallel, the applied stress is shared, and each element is deformed equally. Thus the total stress S is equal to the sum of the viscous stress $\eta\,(d\gamma/dt)$ plus the elastic stress $G\gamma$:

$$S = \eta\frac{d\gamma}{dt} + G\gamma \qquad (5.8)$$

On integration one obtains

$$\gamma = \frac{S}{G}(1 - e^{-tG/\eta}) \qquad (5.9)$$

where the retardation time η/G is equal to the time for the stress to decrease to $1 - 1/e$ of the original value $(1 - 1/e = 1 - 1/2.7 = 0.63)$.

Several Maxwell units arranged in parallel may be used to represent rheological behavior, and several Voigt-Kelvin units arranged in series may be used to represent creep. A series of springs (ladder network) has also been used to represent polymer behavior. In any case, it is not possible to represent polymer behavior with simple Maxwell or Voigt-Kelvin models. Although such models leave much to be desired, they are able to mimic the type and relative importance of certain general types of behavior on a molecular scale.

In a more practical way, Carswell and Nason have classified polymers on the basis of modulus into five categories, as shown in Figure 5.6: (*a*) soft

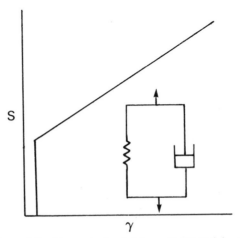

Figure 5.5   Stress–strain plot for a Voigt-Kelvin model.

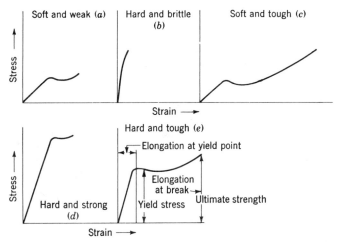

Figure 5.6   Typical stress–strain curves.

and weak, such as polyisobutylene (PIB), (*b*) hard and brittle, such as PS, (*c*) low modulus and high elongation, such as plasticized PVC, (*d*) hard and strong, such as unplasticized PVC, and (*e*) hard and tough, such as ABS (copolymer of acrylonitrile, butadiene, and styrene) and many engineering plastics.

It is important to note that the behavior of polymers below the yield point is Hookean and essentially reversible for short-term service. Thus this range, which is associated with stretching and bending of covalent bonds, is called the *elastic range*. The area under the stress–strain curve is a measure of toughness.

5.9   Viscoelasticity

The early investigations of viscoelasticity using silk, glass, and rubber may be extended to fibers, plastics, and elastomers, respectively. One of the early observations was that these materials undergo very slow irreversible flow or creep when subjected to stress over a long period of time. This irreversible effect is responsible for the so-called flat spots in nylon 66-reinforced pneumatic tires.

Polymers may also fail when subjected to cyclic loading. It is customary to represent these fatigue data by plotting the log of the number of alternating loads causing failure, *N*, against the maximum stress level *S* in the specimen. The *endurance strength* or limit is the stress below which no failure occurs regardless of the number of cycles.

5.10   References

C. D. Armeniades and E. Baer, Chapter 6 in *Introduction to Polymer Science and Technology*, H. S. Kaufman and J. J. Falcetta, eds., Wiley-Interscience, New York (1977).

J. Brandrup and E. H. Immergut, *Polymer Handbook*, Chapter 3, Wiley-Interscience, New York (1976).

F. Bueche, *Physical Properties of Polymers*, Wiley-Interscience, New York (1962).

J. A. Brydson, *Plastics Materials*, Illiffe, Princeton, N.J. (1966); Butterworth (1982).

J. M. Cowie, *Polymers: Chemistry and Physics of Modern Materials*, Intext, New York (1973).

R. D. Deanin, *Polymer Structure Properties and Applications*, Cahners, Boston, Mass. (1972).

G. F. L. Ehlers, Chapter 6 in *Engineering Design for Plastics*, E. Baer, ed., Reinhold, New York (1964).

D. Ferry, *Viscoelastic Properties of Polymers*, Wiley, New York (1960).

A. H. Frazer, *High Temperature Polymers*, Wiley-Interscience, New York (1968).

L. Holliday, Chapter 7 in *Structure and Properties of Oriented Polymers*, I. M. Ward, ed., Wiley, New York (1975).

F. Meares, *Polymer Structure and Bulk Properties*, Van Nostrand, New York (1965).

M. L. Miller, *The Structure of Polymers*, Wiley-Interscience, New York (1964).

L. E. Nielsen, *Mechanical Properties of Polymers and Composits*, Dekker, New York (1974).

R. B. Seymour and C. E. Carraher, *Polymer Chemistry: An Introduction*, Dekker, New York (1981).

M. P. Stevens, *Polymer Chemistry: An Introduction*, Chapter 3, Addison-Wesley, Reading, Mass. (1975).

L. R. G. Treloar, *Introduction to Polymer Science*, Wykeham, London (1970).

P. W. Van Krevelin, *Properties of Polymers*, Elsevier, New York (1972).

D. J. Williams, *Polymer Science and Engineering*, Prentice-Hall, Englewood Cliffs, N.J. (1974).

H. L. Williams, *Polymer Engineering*, Elsevier, New York (1975).

6 | Electric Properties of Polymers

6.1 Dielectric Properties—Contributions

Some of the more important dielectric properties are dielectric loss, loss factor, dielectric constant (or specific inductive capacity), dc conductivity, ac conductivity, and electric breakdown strength. The term *dielectric behavior* usually refers to the variation of these properties within materials as a function of frequency, composition, voltage, pressure, and temperature.

The dielectric behavior is reflected by charging or polarization currents. Since polarization currents depend on the applied voltage and the dimensions of the condenser, it is customary to eliminate this dependence by dividing the charge Q by the voltage V to obtain a parameter C, called the capacity,

$$C = Q/V \tag{6.1}$$

and then using the dielectric constant ϵ, which is defined as

$$\epsilon = C/C_0 \tag{6.2}$$

where C is the capacity of the condenser when the dielectric material is placed between its plates and C_0 is the capacity of the condenser in a vacuum.

Dielectric polarization is the polarized condition in a dielectric resulting from an applied field, either ac or dc. The polarizability is the electric moment per unit volume induced by an applied field of unit effective intensity. The molar polarizability is a measure of the polarizability per molar volume, thus it is related to the polarizability of the individual molecule or polymer repeating unit.

Conductivity is a measure of the number of ions per cubic unit and their average velocity in the direction of a unit applied field. Polarizability is a measure of the number of bound charged particles per cubic unit and their average displacement in the direction of the applied field.

There are two types of charging currents and condenser charges, which may be described as rapidly-forming or instantaneous polarizations, and slowly-forming or absorptive polarizations. The total polarizability of the dielectric is the sum of contributions due to several types of displacement of charge produced in the material by the applied field. The relaxation time is the time required for a polarization to form or disappear. The magnitude of the polarizability, k, of a dielectric is related to the dielectric constant ϵ as shown by the following equation:

$$k = 3(\epsilon - 1)/4\pi(\epsilon + 2) \tag{6.3}$$

The terms "*polarizability* constant" and "*dielectric* constant" can be utilized interchangeably in the qualitative discussion of the magnitude of the dielectric constant. The k values obtained utilizing dc and low-frequency measurements are a summation of electronic E, atomic A, dipole P_0, and interfacial I, polarizations. Only the contribution by electronic polarizations is evident at high frequencies. The variation of dielectric constant with frequency for a material having interfacial, dipole, atomic, and electronic polarization contributions is shown in Figure 6.1.

Instantaneous polarization occurs when rapid (less than 10^{-10} s) transitions take place, i.e., at frequencies greater than 10^{10} Hz or at wavelengths less than 1 cm. Electronic polarization falls within this category and is due to the displacement of charges within the atoms. Electronic polarization is directly proportional to the number of bound electrons in a unit volume and inversely proportional to the forces binding these electrons to the nuclei of the atoms.

The electronic polarization is so rapid that there is no observable effect of time or frequency on the dielectric constant until frequencies are reached which correspond to the visible and ultraviolet spectra. For convenience, the frequency range of the infrared through the ultraviolet region is called the optical frequency range, whereas that including the radio and the audio range is called the electric frequency range.

The electronic polarization is an additive property of atomic bonds, such as C-H, C-C, and C-F. Thus the electronic polarizations and related properties per unit volume are similar for both small molecules and macromolecules.

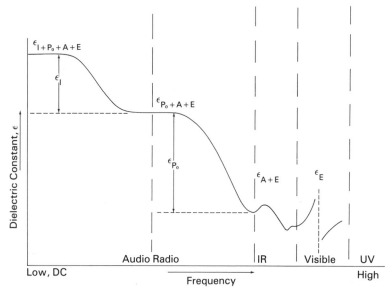

Figure 6.1. Variation of dielectric constant E with frequency for a material having interfacial I, dipole P_0, atomic A, and electronic E polarization contributions.

Accordingly, values obtained for model or small molecules are appropriately applied to analogous polymeric materials. This does not apply in cases where the polymeric nature of the material plays an additional role in the conductance of electric charges, as is the case for whole chain resonance electric conductance.

Atomic polarization is attributed to the relative motion of atoms in the molecule effected by perturbation by the applied field of the vibrations of atoms and ions having a characteristic resonance frequency in the infrared region. The atomic polarization is large in inorganic materials which contain low-energy-conductive bonds and approaches zero for nonpolar organic polymers, such as polyethylene.

The atomic polarization is rapid, and this and the electronic polarizations constitute the instantaneous polarization components. The remaining types of polarization are absorptive types with characteristic relaxation times corresponding to relaxation frequencies.

In 1912, Debye suggested that the high dielectric constants of water, ethanol, and other highly polar molecules are due to the presence of permanent dipoles within each individual molecule. There is a tendency for the molecules to align themselves with their dipole axes in the direction of the applied electric field.

The major contributions to dipole polarizations are additive and are similar whether the moiety is within a small or a large (polymeric) molecule. Even so, the secondary contributions to the overall dipole polarization of a sample are dependent on both the chemical and the physical environment of the specific dipole unit and on the size and the mobility of that unit. Thus dipole contributions can be utilized to measure glass transition temperature T_g and melting point T_m.

These polarizations are the major types found in homogeneous materials. Other types of polarization, called interfacial polarizations, are the result of heterogeneity. Ceramics, polymers containing additives, and paper are considered to be electrically heterogeneous. There are numerous causes of interfacial polarization contributions, and each has its own special theory to explain the particular contribution.

6.2 Delocalization

The probable mechanism for conductance in selected polymers is related to the difficulty of delocalizing electrons. Carraher and colleagues found that antimony polyesters derived from aliphatic diacids are weak semiconductors with bulk specific resistivities in the range of 10^{-10} ohm · cm, whereas analogous polyesters derived from aromatic diacids are good semiconductors with bulk resistivities of about 10^{-4} ohm · cm. The difference in electron conductivity is believed to be due to a difference in ability of the material to delocalize electrons; the aromatic polyesters possess the ability to delocalize electrons throughout the polymer chain and thus are more conductive (lower resistivities).

Delocalization of electrons can be pictured by utilizing either the resonance (valence bond, VB) or assumed delocalization (molecular orbital, MO) concepts. This is illustrated in Figure 6.2 for a two-unit (biphenylene) repeating segment of polyphenylene.

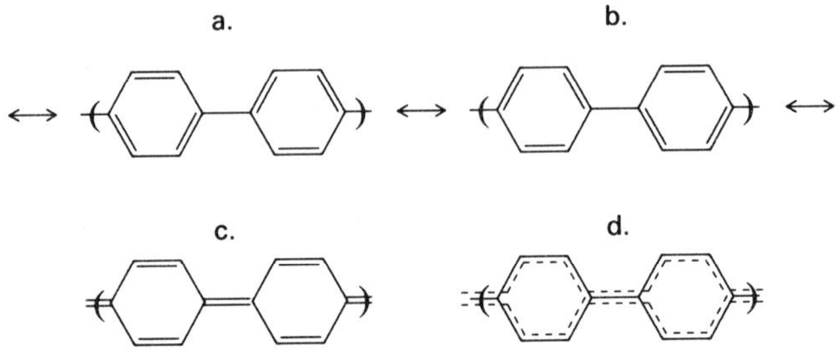

Figure 6.2. Resonance (a,b,c) and MO (d) forms for poly-*para*-phenylene.

Delocalization must extend beyond single chains to allow electrons to travel through pellets having a thickness of several millimeters. Thus polymers with electrons which are readily delocalized are better electric conductors. There is often an interesting interrelation between electron delocalization and color. Color increases both in occurrence and darkening as delocalization increases. The probability that a material will be colored increases as electron delocalization increases. Further, the color of a material typically darkens as the extent of delocalization increases, eventually yielding black products which exhibit good electron delocalization such as the pyrolysis products of PAN and graphite products.

6.3   Direct Current Measurements

A number of factors have been investigated as a response to an imposed (dc) electric field. Current strength, specimen shape, time of measurement, applied pressure, and temperature are typical factors which should be considered.

The bulk (or volume) specific resistance p is one of the most useful electric properties that can be measured. Specific resistance is a physical quantity that may differ by more than 10^{23} in readily available materials. This unusually wide range of conductivity is basic to the wide use of electricity and many electric devices. Conductive materials, such as copper, have p values of about 10^{-6} ohm · cm, while good insulators, such as polytetrafluoroethylene (PTFE) and low-density polyethylene (LDPE), have p values of about 10^{17} ohm · cm.

Specific resistance is calculated from Equation 6.4, in which R is the resistance in ohms, a is the pellet area in square centimeters, t is the pellet thickness in centimeters, and p is the specific resistance in ohm·centimeters.

$$p = R(a/t) \qquad\qquad (6.4)$$

No steady current flows in a perfect insulator in a static electric field, but energy is stored in the sample as a result of dielectric polarization. Thus the insulator acts as a battery which stores energy. In actuality some leakage of current does occur even for the best real insulators.

The insulating property of materials breaks down in strong fields. This breakdown strength, called the electric or dielectric strength (DS), i.e., the voltage required for failure, is inversely related to the thickness l of the material:

$$DS \; \alpha \; l^{-0.4} \tag{6.5}$$

The DS is high for many insulating polymers and may be as high as 10^3 mV/ m. The upper limit of the DS of a material is dependent on the ionization energy present in the material. Electric or intrinsic decomposition (breakdown) occurs when electrons are removed from their associated nuclei; this causes secondary ionization and accelerated breakdown. The DS is reduced by mechanical loading of the specimen and by increasing the temperature.

Breakdown may occur below the measured DS as a result of an accumulation of energy through inexact dissipation of the current; this leads to an increase in temperature and thermal breakdown.

6.4 Dielectric Constant

The dielectric constant (permittivity) [Eq. (6.2)], which is related to the polarizability of the polymer, is low for nonpolar molecules such as high-density polyethylene (HDPE), which cannot store much energy, but is relatively high for polar polymers. The dielectric constant increases as the temperature increases but reaches a plateau at relatively high temperatures.

6.5 Alternating Current

The electric properties of a material vary with the frequency of the applied current. The response of a polymer to an applied current is delayed because of a number of factors including the interaction between polymer chains, the presence within the chain of specific molecular groupings, and effects related to interactions within the specific atoms themselves. A number of parameters are employed as measures of this lag, such as relaxation time, power loss, dissipation factor, and power factor.

The movement of dipoles (related to the dipole polarization, P_0) within a polymer can be divided into two types, an orientation polarization, P'_0, and a dislocating or induced polarization.

The relaxation time required for the charge movement of electronic polarization E to reach equilibrium is extremely short ($\cong 10^{-15}$ s), and this type of polarization is related to the square of the index of refraction, n^2. The relaxation time for atomic polarization A, is about 10^{-3} s. The relaxation time for induced orientation polarization P'_0 is dependent on molecular structure and is temperature-dependent.

A qualitative relationship exists between the electric and mechanical relaxation times. Both are increased by the addition of fillers and reduced by the addition of plasticizers.

The electric properties of polymers are related to their mechanical behavior. The dielectric constant and dielectric loss factor are analogous to the elastic compliance and mechanical loss factor. Electric resistivity is analogous to viscosity.

Polar polymers, such as ionomers, possess permanent dipole moments. Hence orientation polarization is produced in addition to the induced polarization when the polar polymers are placed in an electric field. Thus polar molecules are capable of storing more electric energy than nonpolar polymers, which are dependent almost entirely on induced dipoles for electric energy storage.

The induced dipole moment of a polymer in an electric field is proportional to the strength of the field, and the proportionality constant is related to the polarizability of the atoms in the polymer. The dielectric properties of polymers are affected adversely by the presence of moisture, and this effect is greater in hydrophilic than in hydrophobic polymers.

As shown by the Clausius–Mosette equation,

$$P = \left(\frac{\epsilon - 1}{\epsilon + 2}\right)\frac{M}{\rho} \tag{6.6}$$

the polarization P of a polymer in an electric field is related to the dielectric constant ϵ, the molecular weight M, and the density ρ.

At low frequencies, the dipole moments of polymers are able to keep in phase with changes in a strong electric field, and the power losses are low. However, as the frequency is increased, the dipole moment orientation may not occur rapidly enough to maintain the dipole in phase with the electric field.

The dielectric constant is independent of the frequency at low-to-moderate frequencies but is dependent on the frequency at high frequencies. The dielectric constant is equal to the square of the index of refraction n^2 and to one-third the solubility parameter δ (see Sec. 8.1).

The pendant groups may oscillate in an electric field at temperatures below the T_g, but the polymer chain does not oscillate below the T_g. It is important to note that the T_g as measured electrically may not coincide with the T_g obtained by other techniques. The energy consumed in aligning the polymer dipoles in an alternating field is called the power loss N, i.e., the loss of energy per second. The power loss is zero for a perfect dielectric.

The ratio of the power loss to the power output is called the dissipation factor, tan δ. The power loss is zero when the current and voltage are in phase but increases when the phase difference between the alternating potential and the amplitude of the current is 90°.

6.6   General

Since most polymers consist of covalently bonded catenated carbon atoms, they are nonconductors of heat and electricity. This property is essential when these polymers are used as electric insulators, but is a nuisance when the stored electrostatic charges collect dust or cause electromagnetic interference (EMI).

In contrast to the nonconducting polymers, such as HDPE, polysulfur nitride $(SN)_n$ is a conductor of electricity at ordinary temperatures, and this property is enhanced as the temperature is lowered. The polymer $(SN)_n$ is an anisotropic superconductor at 0.3 K. This conductivity is related to a trans planar conformation of chains with delocalized π orbitals.

When heated, polyvinyl chloride (PVC) and polyvinyl alcohol (PVA) lose HCl and H_2O, respectively, to produce dark-colored conductive polyacetylene. Superior polymers of acetylene can be made by the polymerization of acetylene with Ziegler-Natta catalysts. The conductivity of polyacetylene is increased by the addition of dopants, such as arsenic pentafluoride or sodium naphthenide.

Polymers may also be converted from nonconductors to conductors of electricity by the addition of conductive fillers. Graphite-filled polymers are semiconductors, and polymers filled with aluminum flakes or aluminum filaments are relatively good conductors of electricity.

The Allied Corporation and the University of Pennsylvania have established a joint venture to produce lightweight rechargeable plastic storage batteries based on polyacetylene. MacDiarmid and co-workers at the University of Pennsylvania have produced a battery by immersing polyacetylene film in a propylene carbonate solution of lithium perchlorate. The composition of the film in the charged battery corresponds to $\{CH(ClO_4)_{0.06}\}_n$. This battery will retain its charge over a long period of time.

Since electric charges may be transferred by rubbing surfaces of nonconductors against each other, the ancients were able to develop a crude triboelectric series based on amber, glass, and other available nonconductors. The prefix *tribo* is derived from the Greek word *tribein* meaning "to rub."

Since surfaces of polymers become charged during processing and fabrication, they attract dust, etc. This undesirable property is overcome by the addition of antistats to commercial polymers. Many of these additives attract moisture, and this helps dissipate the electrostatic charge.

The half-lives for loss of charge from untreated polymers vary from a fraction of a second for cellophane to over an hour for PVA. These polymers may be positively or negatively charged, and the half-lives of the two differently charged surfaces may vary. For example, the half-life of positively charged PVA is over twice that of negatively charged PVA.

The charge density of a polymer depends on its polarizability. Polymers with conductivities greater than 10^{-8} ohm $\cdot$ cm^{-1} do not become charged when rubbed against a dissimilar surface, such as a metal.

The electrostatic charge on the polymer depends on both the polymer, the material used as the rubber, and the time of rubbing. Test methods for triboelectric properties have been standardized, and triboelectric series for polymers are available.

Nonconductors, such as polystyrene (PS), may also be made to retain an applied electric charge for a period of time. These so-called electrets may be produced by applying an electric field to a polymer at a temperature of about 35 K above its T_g and allowing it to cool below the T_g while still under the influence of the electric field. Slightly higher temperatures ($T_g + 55$ K) are used when the polymer is allowed to flow under pressure while in the electric field. The retained charges, which are positive on one side and negative on the other side of the polymer, may be retained for several months.

The conductivity of PVC and PVA increases when these polymers are heated. The centers of positive and negative electric charges of nonpolar polymers coincide in the absence of an electric field. However, when a polymer is placed in an electric field, the positive and negative charges in the polymer are displaced, and it becomes a temporary dipole.

6.7 References

T. Alfrey and E. F. Gurnee, *Organic Polymers,* Chapter 9, Prentice-Hall, Englewood Cliffs, N.J. (1967).

C. Carraher, *J. Chem. Educ.* **54**(9), 576 (1977).

C. Carraher, J. Sheats, and C. Pittman, eds., *Organometallic Polymers,* Academic Press, New York (1978).

C. Carraher, J. Sheets and C. Pittman, eds., *Advances in Organometallic and Inorganic Polymer Science,* Dekker, New York (1982).

H. G. Elias, *Macromolecules—Structure and Properties,* Vol. 1, Plenum Press, New York (1976).

G. F. Kinney, *Engineering Properties and Applications of Plastics,* Chapter 17, Wiley, New York (1957).

N. G. McCrum, B. E. Read, and G. Williams, *Antistatic and Dielectric Effects in Polymer Solids,* Wiley, New York (1967).

M. L. Miller, *The Structure of Polymers,* Reinhold, New York (1966).

E. J. Murphy and S. O. Morgan, *Bell Syst. Tech. J.* **16,** 493 (1937).

R. B. Seymour, *Conductive Polymers,* Plenum Press, New York (1981).

P. W. Van Krevelin, *Properties of Polymers,* Chapter 12, Elsevier, New York (1972).

7 | Thermal Properties of Polymers

7.1 Glass Transition and Melting Range

Linear amorphous polymers are glasslike at low temperatures and become leathery at temperatures slightly higher than the glass transition temperature (T_g). These leathery polymers become rubbery at slightly higher temperatures, and crystalline polymers melt at the melting point (T_m).

Polymers with a high cross-link density do not melt. A small degree of cross-linking restricts chain mobility and increases the T_g values. Partly crystalline polymers may be flexible at temperatures above the T_g and below the T_m.

Polymers with flexible chains, such as natural rubber (NR), have low T_g values. The T_g is always less than the T_m, and the ratio of T_g to T_m is lower for symmetrical polymers like polyvinylidene fluoride (PVDF) than for those with unsymmetrical repeating units, such as polychlorotrifluoroethylene. Raymond Boyer has proposed a relationship of $T_m = KT_g$, where the constant $K = 2$ for symmetrical and 1,4-asymmetrical chains.

The free volume, i.e., the volume not occupied by the polymer molecules, is similar for polymers at the T_g and increases as the temperature is increased. More-mobile short chains have lower entropy values and hence lower T_m values, while less-mobile stiff chains have higher entropy values and higher T_m values.

The T_g increases as the intermolecular forces in the polymer and the regularity or crystallinity of the polymer chain structure increase. Thus polyvinyl chloride (PVC) has a higher T_g than linear polyethylene (HDPE) because of the presence of dipole–dipole interactions between the chains in PVC.

Likewise, the T_g of isotactic polypropylene (PP) is greater than the T_g of the less regular atactic PP. The T_g of *cis*-polyisoprene (178 K) is lower than that of *trans*-polyisoprene (190 K).

The values of the T_g and the T_m are lowered by the presence of flexibilizing groups in the polymer chain, such as $-(CH_2)_n-$.

The T_g and the T_m of polymers containing aromatic moieties such as phenylene and isopropylidene-bis-4-phenylene (derived from bisphenol A) are high when compared with those of analogous polymers derived from aliphatic-containing reactants. This is the result of both the rigidity of the aromatic portion and the contribution from electron delocalization.

$$-R-\overset{\overset{\textstyle O}{\|}}{C}-O-\langle\bigcirc\rangle- \quad\longleftrightarrow\quad -R-\overset{\overset{\textstyle O}{\|}}{C}-\overset{+}{O}=\langle\bigcirc\rangle\overset{-}{=} \quad\longleftrightarrow\quad \text{etc.}$$

$$\begin{matrix} H & H \\ | & | \\ +C{=}C+ \end{matrix} \quad\longleftrightarrow\quad \begin{matrix} H & H \\ | & | \\ \text{⊦}C{-}C\text{⊣} \end{matrix}$$

Chain stiffening is also related to steric factors. For derivatives of bis-4-phenylenemethane, the T_g and the T_m typically increase in the order $R = CH_2 < C(CH_3)_2 < C\emptyset_2$ (Table 7.1).

$$-\langle\bigcirc\rangle-R-\langle\bigcirc\rangle-$$

Other stiffening groups are $-SO_2-$, $-\overset{\overset{\textstyle HO}{|\,\|}}{N}C-$, and $\overset{\overset{\textstyle O}{\|}}{{}_{\diagup}C_{\diagdown}}$, which are present in polyaryl sulfones, aramids, and polyarylether ketone (PEEK).

Polyesters produced by the reaction of a glycol with *o*-phthalic acid have less regularity, are less crystalline, and have lower T_m values than those produced from terephthalic acid. The T_m values of aromatic polyesters are higher than those of the corresponding aliphatic esters. The T_m of aromatic polyesters can be reduced by the insertion of methylene groups in the polymer chain.

Likewise, the T_m values of aromatic polyamides (aramids) are higher than those of the corresponding aliphatic polyamides (nylons). The T_m of monadic and dyadic nylons decreases as the number of methylene groups in the chain increases. Thus the T_m values decrease stepwise as the number of

Table 7.1   Chain-Stiffening Effects of Groups Present in Polymer Backbones

Group	Effect (imparts added)
$-O-$	Flexibility
$+CH_2+_2$	Flexibility
	Stiffness
	Stiffness (less than the previous)
	Stiffness
	Stiffness
	Stiffness
	Stiffness

methylene groups is increased from 3 in nylon 4 to 11 in nylon 12. The T_m of monadic nylons with odd numbers of methylene groups is less than that of monadic nylons with even numbers. The T_m of nylon 8 is 473 K, and the T_m of nylon 9 is 482 K. There are eight methylene groups in nylon 9, and they fit more tightly than the seven methylene groups in nylon 8.

Likewise, the T_m is lower when an odd number of methylene groups are present in dyadic nylon. The T_m of nylon 66 is 538 K, while the T_m of nylon 56 is 496 K (Table 7.2).

Regularly arranged bulky pendant groups, such as phenyl groups, increase T_g values. Thus polystyrene has a T_g of 373 K, while that of polyethylene is 153 K. However, flexible, less bulky pendant groups, such as the alkyl ester groups in polyacrylates, decrease T_g values until the groups are large enough to permit side chain crystallization. The T_g is reduced from 338 to 300 K as one goes from polyethyl to polyoctadecyl methacrylate (Table 7.2).

Table 7.2   Factors Affecting T_g Values

Structural variables	Influence on T_g
Increased symmetry	Increase
Increased length of flexible side, pendant groups	Decrease
Inclusion of chain-stiffening units	Increase
Addition of α and α,α substituents	Increase
Addition of polar groups	Increase
Increase in cross-link density	Increase

This is also true of vinyl alkyl ethers. Thus the T_g of polyethyl vinyl ether is lower (244 K) than the T_g of polymethyl vinyl ether (263 K). However, branched alkyl groups increase the T_g, mainly through imparted steric restrictions. Hence the T_g values of polyvinyl n-butyl ether, polyvinyl isobutyl ether, and polyvinyl $tert$-butyl ether are 221, 255, and 361 K, respectively.

Substituents on the phenyl ring in polystyrene (PS) also increase the T_g. Two substituents may have a greater effect on the T_g than one. The T_g values of o-methyl and p-chlorophenyl pendant groups in polystyrene are 388 and 401 K, respectively.

In general, regularity in polymer structure and strong intermolecular forces favor high T_m values. Phenylene groups, such as those in polyphenylene, poly-p-phenylene oxide (PPO), and poly-p-phenylene sulfide (PPS), increase T_m values.

Ladder polymers, such as black nylon, produced by heating polyacrylonitrile, have extremely high T_m values.

Most general purpose linear polymers, such as polyolefins, PS, PVC, and polymethyl methacrylate (PMMA), are not suitable for use at temperatures above 100 °C. PMMA and other polymers of 1,1-substituted vinyl monomers, such as poly-α-methylstyrene, decompose almost quantitatively to their monomers at elevated temperatures. However, the T_g and T_m values of these polymers are greater than those of polymers from 1-substituted vinyl monomers. For example, the T_g values of polymethyl acrylate (PMA) and PMMA are 276 and 381 K, respectively.

7.2   Copolymers

Because of irregularity in structure, the T_m of a nylon 66/610 copolymer is lower than the T_m of either of the homopolymers. The T_g values of these copolymers are also decreased to some extent, but since the T_g is less dependent than the T_m on the packing efficiency of the chain, the effect is much less.

The relation between the T_m and the mole fraction W_A of comonomer A in a copolymer AB is shown by the following expression:

$$\frac{1}{T_{m,AB}} - \frac{1}{T_{m,A}} = -\frac{R}{\Delta H} \ln W_A \qquad (7.1)$$

where $T_{m,AB} = T_m$ of the copolymer; $T_{m,A} = T_m$ of polymer A; ΔH = enthalpy of polymer A; and R = gas constant.

The relation of T_g of many amorphous random copolymers $(AB)_n$ is described by the following expression, where W_A and W_B are the mole fractions of comonomers A and B present in the copolymer $(AB)_n$:

$$\frac{1}{T_{g,AB}} \propto \frac{W_A}{T_{g,A}} + \frac{W_B}{T_{g,B}} \quad \text{and} \quad K_1 W_A(T_g - T_{g,A}) + K_2 W_B(T_g - T_{g,B}) = 0 \quad (7.2)$$

where K_1 and K_2 are constants dependent on the monomers involved.

Each domain in a block copolymer exhibits its characteristic T_g and T_m. Thus the triblock of styrene–butadiene–styrene (Kraton) has a T_g of 373 K for the styrene block and a T_g of 210 K for the butadiene block. In the temperature range of 210 to 373 K, the block copolymer has both high-resilience and low-creep characteristics. The copolymer is rubbery and flows at temperatures above 373 K.

7.3 Additives

The presence of compatible high-boiling liquids (plasticizers) lowers the T_g of polymers such as PVC. The extent of lowering the T_g of the polymer by the addition of W_1 (mass fraction) of plasticizer with a T_g of T_g' is shown by the following expression, in which $T_{g,M}$ is the T_g of the plasticized polymer [directly analogous to Eq. (7.2)]:

$$\frac{1}{T_{g,M}} \propto \frac{W}{T_g} + \frac{W_1}{T_g'} \qquad (7.3)$$

Thus the presence of 40 mass percent of dioctyl phthalate (DOP) plasticizer to PVC lowers the T_g from 395 to 270 K, and hence the plasticized PVC is flexible at room temperature. Solvents have similar effects on the T_g, but of course a volatile solvent is much less permanent than a high-boiling plasticizer.

The T_g of elastomers must be below the use temperature. The high degree of cold flow which is characteristic of polymers at temperatures above the T_g is reduced by the incorporation of a few crosslinks to produce a network polymer with a low crosslink density.

7.4 Thermal Conductivity

As energy—heat, magnetic, or electric—is applied to one side of a material, the energy is transmitted to other areas of the sample. Heat energy is largely transmitted through the increased amplitude of molecular vibrations. The heat flow Q from any point in a solid is related to the temperature gradient dt/dl through the thermal conductivity λ as follows:

$$Q = -\lambda(dt/dl) \tag{7.4}$$

The transmission of heat is favored by the presence of ordered crystalline lattices and covalently bonded atoms. Thus graphite, quartz, and diamond are good thermal conductors, while less-ordered forms of quartz such as glass have lower thermal conductivities. Table 7.3 contains a brief listing of thermal conductivities for a number of materials. Most polymeric materials have λ values between 10^{-1} and 10^{0} W·m^{-1}·K^{-1}.

Crystalline polymers such as high-density polyethylene (HDPE), PP, PTFE, and polyoxymethylene (POM) exhibit somewhat higher λ values than amorphous polymers such as low-density polyethylene (LDPE), atactic PS,

Table 7.3   Thermal Conductivities of Selected Solids

Material	Thermal conductivity (W·m^{-1}·K^{-1})	Material	Thermal conductivity (W·m^{-1}·K^{-1})
Copper	~7200	PS	0.16
Graphite	~150	PS (foam)	~0.04
Iron	~90	PVC	0.16
Diamond	~30	PVC (foam)	~0.03
Quartz	~10	Nylon 66	0.25
Glass	~1	PET	0.14
PMMA	~0.19	NR	0.18
PVC (35% plasticizer)	~0.15	PU	0.31
LDPE	~0.35	PU (foam)	~0.03
HDPE	~0.44	PTFE	0.27
PP	~0.24		

atactic PVC, and atactic PP. The value of λ generally increases with increasing density and crystallinity in the same polymer. For amorphous polymers in which energy can be transmitted through the chain, the conductivity typically increases as the chain length increases. The addition of small molecules such as plasticizers often reduces the thermal conductivity.

Thermal conductivity is not greatly affected by temperature changes as long as the material does not undergo a phase transition. Stretching of polymeric materials producing an anisotropic environment on a molecular scale increases the conductivity along the direction of elongation and decreases it along the shortened axis. The conductivity of HDPE is increased 10-fold along the axis of elongation at $10^3\%$ strain.

Foamed cellular materials have much lower conductivities, typically about 0.03 to 0.07 $W \cdot m^{-1} \cdot K^{-1}$, because gas is a poor conductor. Thus such foams are employed as heat insulators in drinking mugs, commercial insulation, and thermal jugs.

7.5 Heat Capacity

The thermal conductivity is also related to the specific heat capacity C_p as described in Eq. (7.5), where d is the density of the material and TD is the thermal diffusivity:

$$\lambda = (TD)C_p d \tag{7.5}$$

The amount of heat required to raise the temperature of a material is related to the vibrational and rotational motions thermally excited within the sample. Polymers typically have relatively (compared with metals) large specific heats, with most falling within the range of 1 to 2 $kJ \cdot kg^{-1} \cdot K^{-1}$. Replacement of hydrogen atoms by heavier atoms such as fluorine or chlorine leads to lower C_p values. The C_p values change as materials undergo phase changes (such as that at the T_m) but remain constant between such transitions.

7.6 Thermal Linear Expansivity

The thermal linear expansivity of polymers is usually higher than that of ceramics and metals; polymers have values ranging from 4 to 20 $\times$ 10^{-5} K^{-1}, whereas metals have values of about 1 to 3 $\times$ 10^{-5} K^{-1}. Further, the expansion of polymeric materials, unlike the expansion of metals, is usually not a linear function of temperature.

7.7 Thermal Stability

The oxidative stability of the exposed surface of most organic polymers is less than that in inert conditions and within solids. A description of the many known variations of thermal degradation of polymers is beyond the scope of this book.

Within polymer solids, volatile, reactive fragments are trapped and often rereact, forming rearranged structures. If the rearranged structures exhibit markedly better stability, excessive char results. Thus solid LDPE decomposes with little char, whereas polyacrylonitrile (PAN) gives excessive char because of the formation of thermally stable rearranged products.

As with most hydrocarbons, the major products of oxidative degradation are carbon dioxide and water. Most degradation of polymers occurs through a combination of bond scissions and unzipping. Chain scission of PMMA and PTFE begins with the breakage of the chain bond between the terminal unit and the remainder of the chain; this is followed by unzipping. Degradation of PVC initially occurs through evolution of hydrogen chloride:

$$+CH_2-CHCl+ \quad \rightarrow \quad +CH=CH+ \;+\; HCl \;\uparrow$$

Thermal degradation of organic polymers typically begins around 100 °C, and the rate of degradation increases as the temperature increases. Degradation generally shows a somewhat smooth downward slope when weight retention is plotted as a function of temperature after inception of degradation. This occurs because most of the bonds in the polymer chain have similar bond energies and susceptibility to oxidation, since the bulk of the chain is composed of C—C bonds.

A number of techniques, including addition of stabilizers and cross-linking, have been employed to extend the thermal stability of polymers. A number of approaches have been taken to generate polymers that exhibit high-temperature stability. The most common are the incorporation of bonds with greater thermal stabilities (as the $P=N$ bond in polyphosphazenes) and the use of constructive ladder and highly cyclic products, in which one can envision a bond breaking, but being held in place by other bonds until it can reattach itself at its original site. Polybenzimidazoles are examples of synthetic, highly cyclic polymers which exhibit thermal stabilities in excess of 300 °C:

Polyquinoxaline illustrates ladder products:

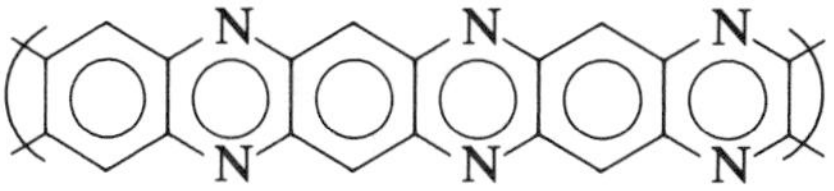

Table 7.4 contains an abbreviated listing of manufacturers' recommended use temperatures for some common materials. Such temperatures must be considered only guides, and the actual use temperatures are subject to actual service conditions and performance requirements.

Polyvinyl acetate (PVAc) loses acetic acid, and polybutyl methacrylate loses butene at high temperatures. Both PVC and polyvinylidene chloride (PVDC) lose chlorine as hydrogen chloride and other halogenated species when heated above 200 °C.

Polyvinyl fluoride and PVDF are more stable to elevated temperatures than the corresponding chloride polymers. The decomposition temperatures of polytrifluoroethylene and polytetrafluoroethylene (PTFE) are progressively higher than polymers of vinyl fluoride or vinylidene fluoride. The pyrolysis of PAN and polymethacrylonitrile yields polycyclic ladder polymers.

General purpose thermosets, such as phenolic (PF), melamine (MF), and epoxy plastics, are more resistant to heat than the general purpose thermoplastics. These crosslinked thermoset polymers do not have true T_m values but decompose at elevated temperatures.

Aromatic cyclic chains are more stable than aliphatic catenated carbon chains at elevated temperatures. Thus linear phenolic and melamine polymers are more stable at elevated temperatures than polyethylene, and the corresponding cross-linked polymers are even more stable. In spite of the presence of an oxygen or a sulfur atom in the backbones of polyphenylene oxide (PPO), polyphenylene sulfide (PPS), and polyphenylene sulfone, these polymers are

Table 7.4　Manufacturers' Recommended Upper Use
Temperatures for Selected Polymeric Materials

Polymer	Short-range use (°C)	Continuous use (°C)
Nylon	180	110
PU	120	80
PTFE	260	—
NR	100	80
PMMA	—	80
EPDM	180	150

stable at relatively high temperatures. Polyimides, such as polybenzimidazole, etc., are at least in part ladder polymers and are also stable at relatively high temperatures.

For inorganic and organometallic polymers, Bailar has suggested some general rules regarding polymer flexibility and thermal stability. Briefly, (1) a metal imparts stiffness to its immediate environment, (2) metal ions stabilize ligands only in their immediate vicinity, (3) thermal, oxidative, and hydrolytic stability are not directly related, (4) metal–ligand bonds have enough ionic character to permit them to rearrange more readily than typical "organic bonds," (5) flexibility increases as the covalent nature of the metal–ligand bond increases, and (6) the coordination number and the stereochemistry of the metal ion determines the polymer structure.

Because of the high bond energy of the siloxane (Si—O) and phosphazene (P = N) groups, siloxane and phosphazene polymers are particularly stable at high temperatures. The T_g values of phosphazene halides increase from 183 K (NPF_2) to 207 K ($NPCl_2$)$_n$ to 265 K ($NPBR_2$)$_n$.

The coefficient of linear expansion of unfilled polymers is approximately 10×10^{-5} cm/cm · K*. These values are reduced by the presence of fillers or reinforcements. The thermal conductivity of the polymers is about 5×10^{-4} cal/sec · cm · K. These values are increased by the incorporation of metal flake fillers. The specific heat is about 0.4 cal/g · K, and these values are slightly lower for crystalline polymers than for amorphous polymers.

The resistance of polymers to flame may be increased by the addition of halogenated compounds and antimony oxide. Organic phosphate additives inhibit the glow of the char formed in burning polymers. Polymers with chlorine pendant groups, such as PVC, and those with halogen-substituted phenyl groups, such as polyesters produced from tetrabromophthalic anhydride, are more flame-resistant than hydrocarbon polymers.

7.8 References

H. R. Allcock and F. W. Lampe *Contemporary Polymer Chemistry,* Part 3, Prentice-Hall, Englewood Cliffs, N.J. (1981).

J. M. G. Cowie, *Polymers: Chemistry and Physics of Modern Materials,* Chapter 14, Intext, New York (1973).

R. Deanin, *Polymer Structure, Properties and Applications,* Cahners, Boston, Mass. (1972).

P. J. Flory, *Principles of Polymer Chemistry,* Cornell University Press, Ithaca, N.Y. (1953).

S. L. Rosen, *Fundamental Principles of Polymer Materials,* Chapter 8, Wiley-Interscience, New York (1982).

* Proper units is simply 1/K. The unit cm/cm·K is used to emphasize the relationship to length.

J. A. Sauer, K. D. Pae, Chapter 7 in *Introduction to Polymer Science and Technology,* H. S. Kaufman and J. J. Falcetta, eds., Wiley-Interscience, New York (1977).

R. B. Seymour, *Modern Plastics Technology,* Chapter 15, Reston Publishing, Reston, Va. (1975).

C. E. Carraher, Thermal characterizations of inorganic and organometallic polymers, *J. Macromol. Sci., Chem.* **A17**(8), 1293–1356 (1982).

P. E. Cassidy, *Thermally Stable Polymers,* Dekker, New York (1980).

A. H. Frazer, *High Temperature Resistant Polymers,* Interscience, New York (1968).

W. W. Wendlandt, *Thermal Methods of Analysis,* Interscience, New York (1964).

8 | Solubility

8.1 Solubility Parameter

The first step in the solution process of a polymeric material by a good solvent is a swelling. Providing the solvent is a good solvent, the intermolecular forces in linear and branched polymers are broken and the polymer dissolves.

The process by which a polymer is dissolved may be described by the Gibbs free energy equation for constant temperature:

$$\Delta G = \Delta H - T\Delta S \qquad (8.1)$$

in which ΔG is the change in free energy, ΔH is the heat of mixing, T is the Kelvin temperature, and ΔS is the change in entropy.

The requirement that ΔG be negative for spontaneous dissolution is readily met if the first term, ΔH, is less than the second term, $-T\Delta S$. Because the polymer chain units are bound through primary, covalent bonds, the units are not free to move independent of their neighboring units. Thus the $-T\Delta S$ term is lower for polymer solution than for the solution of smaller molecules. Most approaches for correlating structure with polymer solubility have focused on the ΔH term. Hildebrand has proposed the following relationship,

$$\Delta H = \phi_1 \, \phi_2 \left[\left(\frac{\Delta E_1}{V_1} \right)^{\!1/2} - \left(\frac{\Delta E_2}{V_2} \right)^{\!1/2} \right]^2 \qquad (8.2)$$

in which ϕ is the partial volume of solvent and solute, and $\Delta E/V$ is the cohesive energy density (the heat of vaporization per unit volume). Hildebrand

95

assigned the symbol δ for $\left(\dfrac{\Delta E}{V}\right)^{1/2}$; the solubility parameter δ has the dimensions
of $(\text{cal}/\text{cm}^3)^{1/2}$ However, the Hildebrand unit (H) is now used in place of this
cumbersome dimension. According to the Hildebrand concept, a polymer
will dissolve if the δ value of the polymer and that of the solvent are similar.
Actually, most linear amorphous polymers will dissolve in solvents having δ
values $\pm$ 1.8 H of that of the polymer.

The δ value is readily calculated from Eq. (8.3), in which ρ = density,
M = molecular weight of the repeating unit, and R is the ideal gas constant.

$$\delta = \left(\frac{\rho(\Delta H - RT)}{M}\right)^{1/2} \tag{8.3}$$

The δ value of a polymer may also be calculated from Small's relationship,
in which ΣG is equal to the summation of the molar attraction constants of
the individual repeating units constituting the polymer:

$$\delta = \frac{D\Sigma G}{M} \tag{8.4}$$

where D is Small's Molar Attraction Constant.

As shown by Eq. (8.5), δ is related to the constant a in the van der
Waals equation, to the surface tension γ, to the index of refraction n, to the
chain stiffness factor M, and to the glass transition temperature (T_g).

$$\delta = \frac{1.2a^{1/2}}{V} = 4.1 \left(\frac{\gamma^{1/3}}{V}\right) 0.43 = 3\eta^2 = M(T_g - 25)^{1/2} \tag{8.5}$$

Crystalline polymers are much less soluble than amorphous polymers at
temperatures below the melting point (T_m). Cross-linked polymers may swell
but will not dissolve.

The solubility of amorphous nonpolar polymers in nonpolar solvents is
readily predictable from the δ values. However, correction must be made
when intermolecular dipole–dipole bonds and hydrogen bonds are present.

The δ values are additive for mixed solvents and for random copolymers.
However, each domain in a block or a graft copolymer exhibits its own
characteristic δ value. Plasticizers are high-boiling solvents, and their selection
can be based on δ values.

8.2 Polymer Compatibility

The compatibility of polymers can be related to their δ values. Thus various polyolefins are compatible, and many of the polyalkyl acrylates and alkyl methacrylates are also compatible with each other (see Tables 8.1, 8.2).

Other compatible commercial systems are as follows: polystyrene (PS) and polyphenylene oxide (PPO); polyvinyl chloride (PVC) and nylon 66; PVC and acrylonitrile–butadiene rubber (NBR); and PS and polycarbonate (PC) (up to 60% PC).

Prior to the introduction of the solubility parameter (solpar) concept, paint chemists used Kauri butanol values, mixed aniline points, and heptane numbers to predict the solubility of resins in aliphatic solvents. These parameters have been replaced, to a large extent, by solpars, but heptane numbers are still used, and these empirical parameters can be converted to solpar values.

Actually, the more useful Hildebrand solpar system is a measure of polarity. For nonpolar solvents and polymers, such as pentane (7.0 H), polytetrafluororethylene (PTFE) (6.2 H), and silicones ($\cong$ 7 H), the solpar values increase as one goes up the homologous series, as shown in Fig. 8.1. Thus the solpar values for heptane, octane, and decane are 7.4, 7.6, and 7.7 H, respectively.

In contrast to the increasing solpar values of aliphatic hydrocarbon solvents, the solubility parameters of polar solvents, such as alkyl halides,

Table 8.1 Solubility Parameters of Resinous Products

	Solubility parameter (Hildebrand's, δ)
Polytetrafluorethylene (PTFE)	6.2
Polydimethylsilicone	7.3
Polyethylene	8.0
Silicone DC-23	8.2
Polymethyl methacrylate (PMMA)	9.2
Polystyrene (PS)	9.3
Alkyd (medium)	9.4
Polyvinyl chloride (PVC)	9.5
Polyurethane (PU)	10.9
Polyvinylidene chloride (PVDC)	12.2
Polyvinyl alcohol (PVA)	12.6
Cellulose acetate	13.6
Polyacrylonitrile (PAN)	15.4

Table 8.2 Solubility Parameters of
Hydrocarbon Solvents

Solvent	Hildebrand's (δ)
Ethane	6.0
Propane	6.4
n-Butane	6.8
n-Pentane	7.0
n-Hexane	7.3
n-Heptane	7.4
n-Octane	7.6
n-Nonane	7.6
n-Decane	7.7
n-Hexadecane	8.0
Isobutane	6.4
Isopentane	6.7
Isohexane	6.8
Isoheptane	6.8
Isooctane	6.9
Neopentane	6.2
Amylene	6.9
Hexene-1	7.4
Shell sol 72	7.2
Shell TS 28	7.4
Varsol No. 2	7.6
VM + P naphtha	7.6
Kerosene	7.2
Turpentine	8.1
Dipentene	8.5
Cyclohexane	8.2
Decalin	8.8
Methylcyclohexane	7.8
Benzene	9.2
Toluene	8.9
Xylene	8.8
Cumene	8.5
Mesitylene	8.8
p-Cumene	8.2
Tetralin	9.5
Solvesso 150	8.5
Naphthalene	9.9

decrease as one goes up the homologous series. Since the solpar value is related to polarity, one should expect a decrease in this value with increasing chain length; the solvents become less polar as the result of an increase in the number of methylene groups, while the number of polar groups remains unchanged.

As shown in Figures 8.2 and 8.3, the solpar values increase with the

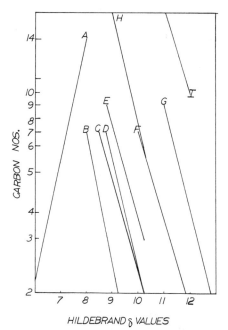

Figure 8.1. Relation of solubility parameters (solpars or Hildebrand δ values) and carbon numbers in various homologous series of solvents. (*A*) Normal alkanes, (*B*) normal chloroalkanes, (*C*) methyl esters, (*D*) alkyl formates and acetates, (*E*) methyl ketones, (*F*) alkyl nitriles, (*G*) normal alkanols, (*H*) alkyl benzenes, and (*I*) dialkyl phthalates.

boiling point T_b, index of refraction η, and density D of solvents in a homologous series. The logarithm of δ values of strongly hydrogen bonded solvents may be estimated from line *C* in Figure 8.4 or from Eq. (8.6a):

$$\log \delta \simeq 0.40 \log \left(\frac{T_b D}{M}\right) \qquad (8.6a)$$

where M is the molecular weight. A constant 0.665 must be added to this expression for moderately hydrogen bonded solvents. The equation for nonpolar solvents is slightly different:

$$\log \delta \cong 0.50 \log \left(\frac{T_b D}{M}\right) + 0.655 \qquad (8.6b)$$

The constant added in this equation is 0.670 for organic solids.

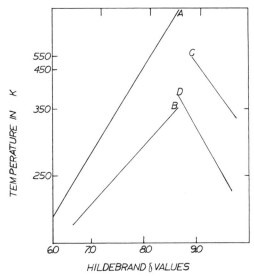

Figure 8.2. Relation of solubility parameters (solpars or Hildebrand values) to boiling points and flash points, where A = boiling points of aliphatic hydrocarbons; B = flash points of aliphatic hydrocarbons; C = boiling points of aromatic hydrocarbons; D = flash points of aromatic hydrocarbons.

Porter has shown for several solvents a correlation between the cohesive energy density (CED) and the Onsager reaction field parameter $(g^2)\left(\dfrac{\eta^2 - 1}{2\eta^2 + 1}\right)^2$, where g is equal to δ (Figures 8.5, 8.6)

8.3 Viscosity

Viscosity is relatively higher in good solvents and lower in poor solvents. Polyisobutylene is used as an additive for motor oils. The oil is a poor solvent for the polymer at room temperature but becomes a good solvent at the operating temperatures of the gasoline combustion engine. Thus although the viscosity of an untreated 10–40 motor oil decreases as the temperature increases, the decrease is counteracted when polyisobutylene is present. Only traces (1–2%) of polyisobutylene are required to achieve the desired viscosity.

Branched chains occupy more volume than linear chains, and the viscosity of polymers with branched chains is less than that of those with linear chains. The viscosities of polymer solutions are greater than those of solvents.

The intrinsic viscosity $[\eta]$ is a measure of the contribution of individual polymer molecules to viscosity. The value $[\eta]$ is related to the size and the

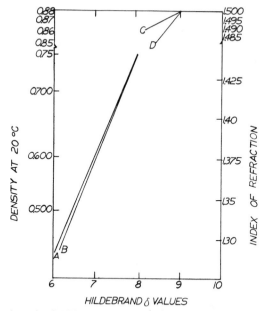

Figure 8.3 Relation of solubility parameters (solpars or Hildebrand values) to index of refraction and density, where A = density of aliphatic hydrocarbons; B = index of refraction of aliphatic hydrocarbons; C = index of refraction of aromatic hydrocarbons; D = density of aromatic hydrocarbons.

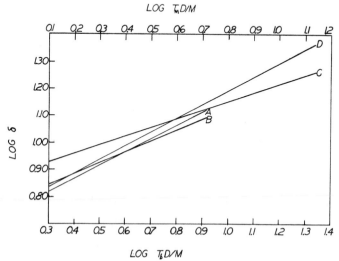

Figure 8.4 Relation of log δ and log of readily available constants, where A = nonpolar liquids; B = moderately polar liquids; C = H-bonded liquids; D = solids.

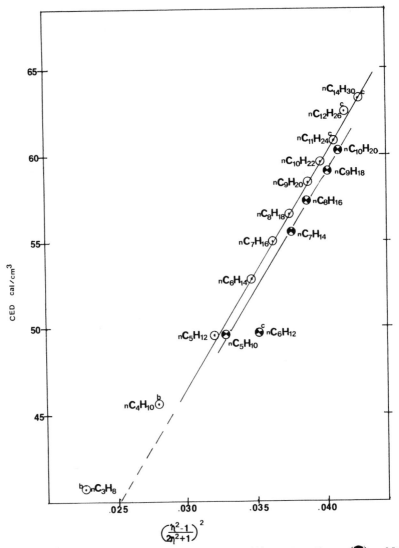

Figure 8.5. Change of CED with g^2 for n-alkanes (⊙) and 1-n-alkanes (⬤) at 25 °C.

shape of the polymer molecules, i.e., to the volume of the polymer chain, as well as to δ values of polymer (δ) and solvent (δ_0) as shown in Eq. (8.7):

$$[\eta] = \eta_0 e^{-\mathcal{H}(\delta \,-\, \delta_0)^2} \qquad\qquad (8.7)$$

The relation between the specific viscosity (η_{sp}) of a dilute solution and that of a concentrated solution is shown by the Huggins equation,

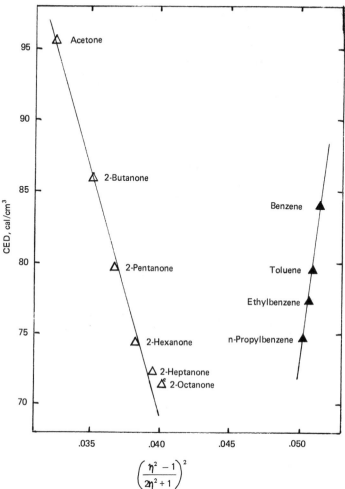

Figure 8.6.   Change of CED with g^2 for methyl-n-alkyl ketones ($\triangle$) and n-alkyl benzenes ($\blacktriangle$) at 25 °C.

$$\eta_{\mathrm{sp}/C} = [\eta] + K'\,[\eta]^2 C \tag{8.8}$$

where the Huggins constant K' is in the range of 0.3 to 0.5, C = concentration, and $[\eta]$ is the limiting viscosity number.

The value of K' in good solvents is typically approximately 0.35 and is largely independent of stereoregularity and molecular weight for high-molecular-weight polymers. However, the value of K' decreases as the molecular weight increases when the molecular weight of the polymer is less than 50,000.

The K' values are usually higher in poor solvents than in good solvents and are usually independent of temperature.

Staudinger showed that $[\eta]$ is proportional to the molecular weight of the dissolved polymer. In the Mark-Houwink equation, this expression has been modified to account for the shape of the polymer chains:

$$[\eta] = KM^a \tag{8.9}$$

Both constants K and a in the Mark-Houwink equation are essentially independent of molecular weight (M) but are dependent on the polymer, the solvent, and the temperature. The constant a is equal to 0.5 in poor solvents in which the polymer molecules are coiled. The constant a is higher, 0.65 to 0.8, in good solvents and may approach 1 for stiff asymmetrical polymer molecules in good solvents. The constant a also increases as the polydispersity of the polymer molecules increases, i.e., the ratio of the weight average molecular weight $\overline{M}_w$ and the number average molecular weight $\overline{M}_n$ increases.

The viscosity of solutions of polymers is inversely related to the temperature. The viscosity may be approximated by the Arrhenius equation at temperatures up to 100 K above the T_g and is more clearly expressed by the Arrhenius equation at higher temperatures:

$$\eta = Ae^{-E_a/RT} \tag{8.10}$$

The activation energy E_a is the energy required for a segment of the polymer to jump into a hole in the solvent. The value of E_a is relatively large (200 to 250 kcal per mole of repeating unit) at temperatures just below the T_g but decreases at temperatures above the T_g. The value of E_a at $T_g + 100$ K is on the order of 5 to 30 kcal per mole of repeating unit.

The solubility of polymers decreases slightly as the molecular weight increases. This difference is sufficient to permit the separation of high-molecular-weight fractions of polymers by the addition of small amounts of a poor solvent to a polymer solution. It is also sufficient to permit preferential extraction of low-molecular-weight fractions of polymers. When a solution of a polymer is cooled, the first fraction to precipitate is the highest-molecular-weight fraction. It is customary to separate a solution of a polydisperse polymer into about five fractions and then to separate each of these fractions into three subfractions.

The solvent precipitation technique is also used to encapsulate materials, such as medicinals and insecticides in polymers. In this technique, the material to be encapsulated is added to the polymer solution before the addition of the nonsolvent.

The coefficient of viscosity η is the proportionality constant when the stress S is related to the viscosity gradient or rate of flow $d\gamma/dt$ in the Newton equation:

$$S = \eta \frac{d\gamma}{dt} \tag{8.11}$$

Polymer solutions that comply with Newton's law for ideal solutions are said to be Newtonian.

Flory has defined the theta temperature as that temperature at which the polymer exists in a solvent in an unperturbed conformation. The polymer chain expands as the temperature is increased and contracts as the temperature is decreased below the theta temperature.

8.4   References

A. F. M. Barton, Solubility parameters, *Chem. Rev.* **75** (73), 1 (1975).

H. Burrell, "Solubility Parameters for Film Formers," presented at meeting of American Chemical Society in Cincinnati, Oh., April 1955.

H. Burrell, Solubility parameters, in *Polymer Handbook*, Chapter 4, J. Brandrup and E. H. Immergut, eds., Wiley-Interscience, New York (1975).

M. B. Djordjevic and R. S. Porter, Chapter 16 in *Macromolecular Solutions*, R. B. Seymour and G. A. Stahl, eds., Pergamon Press, Elmsford, N.Y. (1982).

J. D. Ferry, *Viscoelastic Properties of Polymers*, Wiley-Interscience, New York (1970).

P. J. Flory, *Principles of Polymer Chemistry*, Cornell University Press, Ithaca, N.Y. (1953).

C. M. Hansen, Three dimensional solubility parameters, *Chem. Tech.* **2**, 547 (1972).

F. Harris and R. B. Seymour, *Structure-Solubility Relationships in Polymers*, Academic Press, New York (1977).

J. Hildebrand and R. Scott, *Solubility of Nonelectrolytes*, 3rd ed, Reinhold, New York (1949).

R. B. Seymour, *Solubility Parameters*, Tamkang University Press, Tapei, Taiwan (1980).

R. B. Seymour, Relationships of solubility parameters to molecular weight in a homologous series, *Aust. Paint. J.* **13**(10), 18 (1968).

R. B. Seymour and J. M. Sosa, Relationships of solpar values to density, *Nature* **248**(5451), 759 (1974).

R. B. Seymour and G. A. Stahl, *Macromolecular Solutions*, Pergamon Press, Elmsford, N.Y. (1982).

D. W. Van Krevelen, *Properties of Polymers*, 2nd ed., Chapter 10, Elsevier, New York (1976).

9 | Diffusion and Permeation of Gas and Vapors in Polymers

9.1 Introduction

The diffusion of gases and vapors in polymers and the permeability of polymers to gases and vapors are not only of practical significance in packaging and coatings but may also be used to demonstrate the kinetic agitation of the diffusate and the permeate molecules. The diffusion process, such as the dissolution of a polymer in a solvent; the plasticization of a polymer, such as polyvinyl chloride (PVC); and the passing of gases through films, is dependent on random jumps and hole filling by small diffusate molecules.

The diffusion of larger organic vapor molecules is related to absorption. The rate of diffusion is dependent on the size and shape of the diffusate molecules, their interaction with the polymer molecules, and the size, shape, and stiffness of the polymer chains. The rate of diffusion is related directly to the flexibility of the polymer chain and inversely to the size of the diffusate molecule.

The diffusion and the permeability are inversely related to the density, degree of crystallinity, orientation, filler concentration, and crosslink density of a polymeric film. As a general rule, the presence of plasticizers or residual solvents increases the rate of diffusion in polymers. Films cast from poor solvents have high permeability. The rate of diffusion or permeability is independent of the molecular weight of the polymer, providing the polymer has a moderately high molecular weight.

The permeation of a gas, vapor, or liquid through a polymeric film is a three-step process which involves (1) the rate of dissolution of the low-molecular-weight substances in the polymer, (2) the rate of diffusion of the former through the film in accordance with the concentration gradient, and (3) the energies of the low-molecular-weight substances on the opposite side of the polymer.

Permeation is dependent on the segmental motion of the polymer chains and the free volume of chain segments. The free volume decreases, whereas the chain stiffness increases, as the temperature of the polymeric membrane is lowered toward the glass transition temperature T_g. The free volume is similar for all polymers at the T_g.

The diffusion of a liquid or a gas through a membrane is similar to the diffusion in liquid systems. The mechanism in each case involves a transfer of a small molecule to a hole in the liquid or membrane. For diffusion, the jumping frequency from hole to hole is dependent on the activation energy E_D, which is dependent on the size and shape of the diffusate molecule and the size of the holes in the membrane, i.e., the free volume. For permeation, activation energy E_P is in the order of 5 to 10 Kcal/mol; the values of E_D are somewhat higher.

The rate of diffusion, D, and the rate of permeability, P, increase exponentially as the temperature increases, as shown by the Arrhenius equation for diffusion

$$D = D_0 e^{-E_D/RT}$$

where D_0 is the rate of diffusion at the base temperature, R is the ideal gas constant, and T is the Kelvin temperature.

As shown by Fick's law,

$$F = -D \frac{dc}{dx}$$

the weight of diffusate crossing a unit area per unit time, F, is proportional to the concentration gradient dc/dx. The proportionality constant D is related directly to the pressure differential across the membrane and inversely to the membrane thickness.

When the diffusion coefficient D is dependent on concentration, the diffusion process is said to be Fickian. In such cases, D is related inversely to the solubility S and directly to the permeability P as follows:

$$D = \frac{76P}{S}$$

It has been postulated that linear alkanes diffuse through the holes in the membrane by alignment with the segments of the polymer chains. Such alignments are more difficult for branched alkanes and hence they diffuse more slowly.

The diffusion coefficient D is always related inversely to the crosslink density of vulcanized elastomers. When D is extrapolated to zero concentration of the diffusate, it is related to the weight of the principal section of the elastomer, i.e., the weight of the segments between crosslinks.

At temperatures above the T_g, the diffusion of gases in amorphous polymers may be estimated from the following expression:

$$\log D_0 = 0.5 \frac{(E_D - 8)}{1000}$$

At temperatures below the T_g, the diffusion of gases in amorphous polymers may be estimated from the following equation:

$$\log D_0 = 0.4 \frac{E_D - 10}{1000}$$

The activation energy is lower at temperatures below the T_g than when in the rubbery state, as shown in the following equation:

$$E_{Dg} \cong 0.75 E_{Dr}$$

where E_{Dg} is the activation energy below the T_g and E_{Dr} is the activation energy in the rubbery state.

As shown by the following expression, the diffusion coefficient for crystalline polymers, D_c, is less than that in amorphous polymers, D_a, and is dependent on the extent of crystallinity x.

$$D_c = D_a (1 - x)$$

The solubility of gases in elastomeric membranes is related to the molecular weight of the diffusate molecule. Thus the ratio of the diffusion of oxygen to that of nitrogen differs in natural rubber and in ethylcellulose.

The permeability coefficient P is related to the diffusion coefficient D and the solubility coefficient S as shown by Henry's law:

$$P = DS$$

Thus the permeability values are high when the solubility parameter of the diffusate and that of the polymeric films are similar.

Polymers may exhibit non-Fickian diffusion below the T_g, but a transition to Fickian diffusion is noted as the temperature is raised above the T_g.

The permeability of a gas, such as oxygen, is much greater through a silicone membrane than through a polyvinyl alcohol (PVA) membrane. Permeability of specific gases may be controlled by the use of composite membranes.

The relation between heat of solution, ΔH, and temperature may be calculated from the Clausius-Clapeyron equation:

$$\Delta H = -R \frac{d \ln S}{d(1/T)}$$

Van Amerogen has shown that the heat of solution in the rubbery state, ΔH_r, can be calculated from the following expression:

$$\Delta H_r = -2400 - 2000 \log S(298)$$

Meares has shown that the heat of solution in the glassy state, ΔH_g, is greater than ΔH_r and may be calculated from the following expression:

$$\Delta H_g = 1.67 \Delta H_r - 5000$$

Van Amerogen has also shown that the heat of solution ΔH_f of gases in elastomers may be calculated from the following expression:

$$\Delta H_f = -2400 - 2000 \log S(298)$$

He also showed that the size of the gas molecule is an important factor and that the solubility coefficient S of gases in amorphous polymers can be calculated from the following expression, which relates the critical temperature T_{ct} and boiling point T_b:

$$\log S(298) = -2.1 + 0.0074 T_{ct} \text{ or } -2.1 + 0.0123 T_b$$

The solubility of gases (S_g) in crystalline polymers, S_x, can be estimated from the extent of crystallinity, x_c:

$$S_x(298) = S_g 298(1 - x)$$

The values of D, P, and S of a polymer, such as polyethylene, are a function of the polymer structure and may be altered by the introduction of pendant groups. The values of D and P, and to some extent S, are reduced when a chlorine pendant group is inserted in polyethylene, as in PVC. These values are reduced further by the introduction of a second chlorine pendant group (as in polyvinylidene chloride, PVDC).

The reciprocal of permeability is an additive property. Hence the permeability of many laminates to gases is impeded. Thus it is possible to develop both barrier films and films that may be used for the selective separation of gases and for less permeable plastic containers.

9.2 References

J. Crank and G. C. Park, eds., *Diffusion in Polymers,* Academic Press, New York (1968).

F. Harris, R. B. Seymour, *Solubility-Property Relationships in Polymers,* Academic Press, New York (1975).

J. H. Hildebrand and H. L. Scott, *Solubility of Nonelectrolytes,* Reinhold, New York (1950).

P. Meares, *Polymer Structure and Bulk Properties,* Van Nostrand-Reinhold, New York (1965).

W. R. Moore, *Progress in Polymer Science,* Vol. 1, A. D. Jenkins, ed., Pergamon Press, New York (1967).

R. B. Seymour and C. E. Carraher, *Polymer Chemistry: An Introduction,* Chapter 3, Dekker, New York (1981).

R. B. Seymour, and G. A. Stahl, *Macromolecular Solutions,* Pergamon Press, Elmsford, N.Y. (1982).

V. Stannett and N. Szwarc, *J. Polym. Sci.* **15**, 81 (1955).

S. B. Tuwiner, *Diffusion and Membrane Technology,* Reinhold, New York (1962).

G. J. van Amerogen, *J. Polym. Sci.* **5**, 307 (1950).

R. Week, N. H. Alex, H. L. Frisch, V. Stannett, and M. Szwarc, *Ind. Eng. Chem.* **47**, 2524 (1955).

H. Yasuda, Permeability constants, Section 5 in *Polymer Handbook,* J. Brandrup and E. H. Immergut, eds., Wiley-Interscience, New York (1975).

10 | Chemical Resistance of Polymers

10.1 Rate of Attack

If the chemistry of polymer molecules were different from that of simple compounds resembling the repeating units (model compounds), the study of the chemical resistance of organic polymers would be difficult. Fortunately, Nobel laureate Paul Flory found that the rate of esterification of molecules with terminal hydroxyl and carboxyl groups is essentially independent of the size of the molecules. Thus it is customary to assume that the rates of most reactions of organic molecules are similar regardless of the size of the molecule.

Of course, the rates of reaction are dependent on the accessibility of the functional groups of the polymer to the reactant. This accessibility is enhanced by the presence of solvents and plasticizers and hindered by crystallization, fillers, and cross-links. In general, the rate of attack of a polymer molecule by a corrosive is dependent on the rate of diffusion of the attacking molecule, and the temperature. Polymers are more readily attacked by corrosives at temperatures above the glass transition temperature T_g and when stiffening groups are not present in the polymer molecule.

The resistance of polymers to corrosives is dependent on structure. Some polymers, such as cellulose acetate, do not have outstanding resistance to acids and alkalis. Most polymers are less brittle than the more chemically resistant ceramics and more resistant to corrosives than most metals.

The rate of attack of polymer molecules by corrosives may be enhanced or hindered by the presence of neighboring groups. Carbon bonds with oxygen, sulfur, or nitrogen are more readily cleaved by acids, bases, and other corrosive materials than carbon–carbon bonds since most of these corrosive materials are ionic and/or highly polar in nature.

10.2 Chemical Oxidation

Molecules at surfaces are typically the most susceptible to both chemical and physical attack since at least one side of the polymer chain is unprotected by neighboring segments. Further, the surface has the greatest concentration of defects, dislocations, and sites of chain folding. As surface chains are chemically and/or physically removed, other chain segments are exposed, gradually leading to a degeneration of the polymeric material and associated properties. Many materials are given special surface treatments and modifications in an attempt to slow or stop such degradation.

The effects of solvents on polymeric materials are usually physical rather than chemical in nature. The primary polymer chains remain intact, but the molecular structure is changed, the magnitude of the change typically decreasing as one moves from the material surface (the initial point of solvent–polymer contact) into the bulk portion of the sample.

Polymeric materials are susceptible to solution by solvents or liquids when the solubility parameters (see Secs. 8.1 and 8.2) of the polymer and solvent or liquid are similar. Insertion of mild cross-linking discourages dissolution and is often employed to prevent dissolution of polymeric materials.

Even so, a number of chemical agents, including liquids, chemically attack polymers. Reactions that would ordinarily occur with small molecules also occur in polymers, given the same functional groups and reactive sites. Thus benzene, toluene, etc., are readily sulfonated when exposed to sulfuric acid. Likewise polystyrene (PS) is sulfonated when exposed to liquids and gases containing sulfuric acid:

$$\text{+CH}_2\text{—CH+} + \text{H}_2\text{SO}_4 \longrightarrow \text{+CH}_2\text{—CH+}$$
$$\bigcirc \qquad\qquad\qquad\qquad \bigcirc\text{—SO}_2\text{OH}$$

The common paraffinic polymers, polyethylene (PE) and polypropylene (PP), are relatively inert to chemicals but are attacked by strong chemical reagents. Thus PE can be chlorinated in the same way that paraffins are.

$$\text{+CH}_2\text{—CH}_2\text{+} + \text{Cl}_2 \longrightarrow \text{+CH}_2\text{CHCl+} + \text{HCl}$$

Polymers containing polar groups, such as polyvinyl chloride (PVC), and condensation polymers, such as Nylon 66, are more susceptible to attack than other polymers. Most commerical suppliers of polymeric materials supply data sheets which indicate resistance (and lack thereof) to common solvents.

Organic materials, including polymers, are susceptible to oxidation and ozonization. Thus many materials contain antioxidants to prevent or retard oxidation.

Materials exposed to the out-of-doors are particularly vulnerable since they are exposed to a higher concentration of chemicals (such as SO_3, HCl, and O_3); ultraviolet radiation, which promotes many chemical reactions; and moisture, which can lead to increased crazing and fracture of surface as the seasons change. Further, surface temperatures can reach almost 100 °C on a hot day, increasing the susceptibility to attack as well as the rate of reactions already proceeding.

For vinyl polymers, oxidation is believed to begin by the decomposition of peroxides incorporated as impurities during synthesis and processing. The peroxides decompose, forming free radicals, which in turn abstract hydrogen atoms from adjacent polymer chain segments, which in turn can abstract other hydrogen atoms, leading eventually to the formation of cross-links; this tends to stiffen the material, making it more susceptible to stress cracking. Further, abstraction occurs more readily with tertiary hydrogen atoms

$$\left(CH_2-\underset{\underset{CH_3}{|}}{\overset{\overset{H}{|}}{C}}\right) \leftarrow \text{tertiary hydrogen}$$

(as in PP and PS) than with secondary hydrogen atoms, making PP more susceptible to such oxidation than high-density polyethylene (HDPE), which is also less susceptible to oxidation than low-density polyethylene (LDPE), which has tertiary hydrogen atoms at the juncture of each chain branch. Double bonds activate adjacent hydrogen atoms; thus polybutadiene is more susceptible to oxidation than HDPE.

Oxidation is also dependent on the permeability of the polymer to oxygen. Table 10.1 lists the permeabilities of selected polymers to oxygen. Because bulk oxidations are dependent on the permeability to oxygen, crystalline polymer forms are more resistant to oxidation than amorphous forms. Also, the very nature of the molecules present in the chains affects the tendencies toward oxidation. Thus the fluorine atom in polytetrafluoroethylene (PTFE)

Table 10.1   Oxygen Permeabilities of Selected
Polymers

Polymer	Permeability 10^{15} (kg $\cdot$ m^{-1} $\cdot$ kPa^{-1} $\cdot$ s^{-1})
HDPE	6
LDPE	30
PS	27
PVC	0.5
PAN	0.002
PVA	0.0006
PP	10

resists abstraction compared with hydrogen atoms, and PTFE shows good
resistance to oxidation.

Most materials are susceptible to chemical attack by ozone. Unsaturated
polymers, such as polybutadiene (PB) and natural rubber (NR), are particularly vulnerable to chemical attack by ozone which causes chain scission.

10.3   Stress Cracking and Crazing

Stressing of materials may cause minute cracking in the surface, leaving
a site which is particularly vulnerable to physical and chemical attack. Further,
points of stress are also more susceptible to attack since the molecules present
at the sites of greatest stress are strained. The tendency to stress crack can
be increased when the surface interacts with solvents and chemicals.

Organic liquids and gases can promote the formation of networks of
voids and crazes in amorphous materials. If such crazing can be limited to
the surface, the only major detraction is appearance. Usually such crazing is
a prelude to further cracking and crazing.

10.4   Combustion

The thermal stability of polymeric materials was discussed in Sec. 7.7.
Here we consider briefly additional topics related to burning in an oxygen-
containing atmosphere (air).

In contrast to the much slower environmental oxidation, combustion is
a rapid oxidation process which occurs above a critical, ignition temperature.
Burning in bulk materials is complex, involving at least three environments—
the surface, where oxidation is predominant; the inner surface, where the

amount of oxygen is depleted, yet where mainly gases which are directly evolved are produced; and the bulk, where degradation occurs in an oxygen-free environment, and where gaseous fragments produced are not free to escape. Even in such a complex situation, oxidation (partial or full) eventually occurs, since new surfaces are created as older ones are volatilized; as non-oxidized fragments come into contact with oxygen, they become oxidized.

Most burning is exothermic, consequently feeding on itself. Table 10.2 lists the heats of combustion of selected polymers. It is interesting to note that although cellulosic materials are considered quite flammable, the heat (enthalpy) of combustion of most other polymeric materials is greater. This is because cellulose contains a number of hydroxyl groups, whereas the additional cited polymeric materials contain proportionally less oxygen. Thus cellulose and polyoxymethylene (POM) (which also has a high proportion of oxygen) are already partially oxidized.

The heat of combustion does not indicate the tendency to burn or the rate of burning. The tendency to burn is typically described in terms of the limiting oxygen index (LOI). Briefly, the sample, in a predescribed standard form, is set afire in an upward-flowing oxygen–nitrogen gas mixture, and a stable flame is established. The ratio of oxygen to nitrogen is reduced until the sample flame becomes unstable and is extinguished. The minimum oxygen content which supports combustion is the LOI of the sample. Such tests are quantitative but must be considered first approximations.

Table 10.3 lists the LOIs of select materials. Materials with LOI values above about 0.25 can be considered self-extinguishing under normal atmospheric conditions. Addition of flame retardants can increase the LOI.

Most polymers burn, producing nontoxic combustion materials—CO_2

Table 10.2 Heats of
Combustion of Selected
Polymers

Polymer	ΔH (kJ $\cdot$ g^{-1})
POM	17
Cellulose	18
PVC	20
PET	22
PMMA	26
Nylon 66	32
PS	42
NR	45
PP	46
PE	47

Table 10.3 Limiting Oxygen Indices (LOIs) of
Selected Polymers

Polymer	LOI
PMMA	0.17
PP	0.17
PE	0.17
PE (with 60% $Al_2O_3 \cdot 3H_2O$ added)	0.30
PS	0.18
PC	0.27
PVC	0.48
PTFE	0.95

and H_2O. Some yield some toxic gases. Most polymers yield CO under oxygen-poor burning conditions; nitrogen-containing polymers may produce HCN; and halogenated polymers decompose to form HX and X_2CO gases even in the presence of ample supplies of oxygen.

10.5 Polyolefins

Polyolefins, like their model compounds (alkanes), are unreactive molecules. When ignited in an abundance of oxygen, polyolefins burn and produce carbon dioxide and water. Fluorine reacts with polyolefins at room temperature with explosive violence, but iodine is unreactive even at elevated temperatures.

Chlorine and bromine react with polyolefins at elevated temperatures or in the presence of ultraviolet radiation. The chlorination of polyolefins also takes place in the presence of sulfur dioxide.

The effect of temperature on reactions with polymers can be predicted from the Arrhenius equation, $k = Ae^{-E_a/RT}$.

Polyolefins are resistant to aqueous solutions of inorganic acids such as hydrochloric, phosphoric, and sulfuric acids. These polymers are also resistant to chromic and nitric acids but react with them as well as with dinitrogen tetroxide at elevated temperatures. Polymers with tertiary hydrogen atoms, such as PP, are more readily oxidized than those with only secondary hydrogen atoms, such as HDPE.

It is important to note that although methylene CH_2 groups are considered resistant to typical corrosive agents, they undergo the characteristic reactions of alkanes.

Unsaturated polymers, such as polydienes, react with chlorine and bromine at normal temperatures and in the absence of ultraviolet radiation. The inorganic acids, sulfuric, hydriodic, hydrobromic, hydrochloric, and hydro-fluoric acids, also add to the double bond in unsaturated polymers at normal temperatures. Unsaturated polymers are also less resistant to oxidation than polyolefins. Thus oxidative reactants and ozone may cleave the double bonds in unsaturated polymers.

10.6 Halogenated Aliphatic Polymers

Halogenated aliphatic polymers such as polyvinyl chloride (PVC) and polyvinylidene chloride (PVDC) are moderately resistant to attack by reactants. The fluorinated polymers, such as PTFE, are exceptionally resistant to attack by acids and alkalis even at elevated temperatures because of their tight packing and high C—F bond energy.

10.7 Hydroxyl-Containing Polymers

Polymers with hydroxyl groups undergo typical reactions of alcohols. Those with aromatic hydroxyl groups, such as phenolic resins, are attacked by aqueous alkaline solutions.

Aliphatic polymers with hydroxyl groups, such as cellulose and polyvinyl alcohol (PVA), are attacked by concentrated mineral acids. The principal effect of inorganic acids on cellulose is the cleavage of acetal linkages, producing lower-molecular-weight products. However, there is no decrease in the degree of polymerization as the result of a replacement of a pendant hydroxyl group by an acid anion.

The secondary alcohol groups in PVA may be oxidized to ketones, and the primary alcohol groups in carbohydrates may be oxidized to carboxylic acids. Although these reactions do not reduce the degree of polymerization, they do increase the degree of water solubility of the polymers.

10.8 Condensation Polymers

Esters, such as those present in the pendant groups of acrylic and meth-acrylic esters, may be hydrolyzed by acids or alkalis. The polymeric acids produced are hydrophilic, but the degree of polymerization is unchanged as a result of the hydrolysis. Methacrylic esters are more resistant to hydrolysis than acrylic esters.

Amide, urethane, and ester groups in the polymer chain, such as those present in nylons and polyesters, may be hydrolyzed by acids to produce lower-molecular-weight products. Polyacetals are also degraded by acid hydrolysis, but ethers, such as polyphenylene oxide (PPO), are resistant to attack by acids.

10.9 Aromatic Polymers

Aromatic polymers, such as PS, are readily attacked by chlorine, bromine, concentrated sulfuric acid, and nitric acid. These reactions do not decrease the degree of polymerization of the polymers. Aromatic polymers with stiffening groups, such as PPO, polyarylsulfone, polyarylether ketone (PEEK), and polyphenylene sulfide (PPS), are more resistant to attack by corrosives than those with flexibilizing groups.

10.10 References

R. F. Eisenberg, R. R. Kraybill, and R. B. Seymour, Chapter 1 in *Laboratory Engineering and Manipulation*, 3rd ed. E. S. Perry and A. Weissenberger, eds., Wiley, New York (1979).

C. Hall, *Polymer Materials*, Macmillan, London (1981).

H. F. Mark, Chpt. 2 in *The Effects of Hostile Environments on Coatings and Plastics*, D. Garner and G. A. Stahl, eds., ACS Symposium Series, 229, Washington, D.C., 1983.

R. T. Morrison and R. N. Boyd, *Organic Chemistry*, Allyn and Bacon, Boston, Mass. (1973).

R. B. Seymour and R. H. Steiner, *Plastics for Corrosant Resistant Applications*, Reinhold, New York, (1955).

R. B. Seymour, Part 3, Section 10, in *Treatise on Analytical Chemistry*, I. M. Kolthoff, D. J. Elving, and F. H. Stross, eds., Wiley-Interscience, New York (1976).

R. B. Seymour, *Plastics vs Corrosives*, Wiley, New York (1982).

T. W. Solomon, *Organic Chemistry*, Wiley, New York (1980).

11 | Effect of Additives on Polymers

11.1 Introduction

The properties of polymers may be improved by the presence of appropriately selected additives. With the exception of some cast plastics, such as polymethyl methacrylate (PMMA), and some fibers, such as unpigmented cotton, most commercial polymers are mixtures of the polymer with one or more additives.

The use of additives is not new: Charles Goodyear used carbon black as a pigment for natural rubber; A. F. Critchlow used wood flour as a filler for shellac; and John W. Hyatt used camphor as a flexibilizing agent or plasticizer for cellulose nitrate over a century ago. More recently, hindered phenols have been used as antioxidants for many thermoplastics; glass fibers, treated with alkylsilanes and other coupling agents, have been used as reinforcements for unsaturated polyesters; and other fibers, such as graphite, boron, and aramid, have been used as reinforcements for epoxy resins.

These additives usually enhance specific properties of polymers. Thus solid, pure stereoregular polystyrene (PS) is brittle; yet, as a result of the addition of the proper impact modifiers and other additives, the modified PS exhibits the properties of a good plastic and rubber.

The types of additives and their purposes (the word *additive* is derived from "addition" and simply means "materials added") vary, and the exact proportions and nature of the additives are as much an art as a science. Some of the more important types of additives are discussed in this chapter. It is important to note that the addition of additives often requires additional processing steps. This increases the cost of the finished item; the additives themselves may vary in cost from low-cost clay fillers and sulfur to more expensive biocides.

11.2   Fillers

As shown by the modified Einstein, Guth, and Gould equation,

$$M_c = M_0 (1 + 0.67fC + 1.62f^2C^2) \qquad (11.1)$$

the increase in flexural modulus, or stiffness, of a filled polymer, M_c, is greater than that of the unfilled polymer, M_0, and is related to the concentration C and aspect ratio f (ratio of length to diameter of the filler).

Many properties of composites filled with nonreinforcing fillers, such as coefficient of expansion, heat deflection, and specific heat, may be estimated from the rule of mixtures. Thus the coefficient of expansion of the composite, a_c, is related to the sum of the coefficients of expansion of the continuous phase or resin matrix m and the discontinuous phase or filler f times their fractional volumes V and $(1 - V)$, respectively, as follows:

$$a_c = a_m V + a_f (1 - V) \qquad (11.2)$$

Some common fillers are shown in Table 11.1.

Over 1 million tons each of calcium carbonate and carbon black and over 150 thousand tons of alumina trihydrate (ATH) are used annually as fillers by the U.S. polymer industry. Asbestos continues to be used in moderate amounts (250 thousand tons annually), but it is being displaced by other fillers because of its toxicity.

Many semicompatible rubbery polymers are added to increase the impact resistance of other polymers, such as PS. Other comminuted resins, such as silicones or polyfluorocarbons, are added to increase the lubricity of other plastics. For example, a hot melt dispersion of polytetrafluoroethylene (PTFE) in polyphenylene sulfide (PPS) is used as a coating for antistick cookware.

Carbon black is now the most widely used filler for polymers. Much of the 1.5 million tons produced annually in the United States is used for the reinforcement of elastomers.

Carbon-filled polymers, especially those made from acetylene black, are fair conductors of heat and electricity. Polymers with fair conductivity have

Table 11.1 Fillers for Polymers

I. Organic fillers
  A. Cellulosic products
    1. Wood
    2. Comminuted cellulose
    3. Fibers (cellulose, cotton, jute, rayon)
  B. Lignin-based
  C. Synthetic fibers
    1. Polyesters
    2. Nylon and aramids
    3. Polyacrylonitrile
    4. Polyvinyl alcohol (PVA)
  D. Carbon
    1. Carbon black
    2. Graphite whiskers and filaments
    3. Ground petroleum coke
II. Inorganic Fillers
  A. Silicates
    1. Minerals (asbestos, mica, China clay {kaolinite}, talc)
    2. Synthetic (calcium silicate, aluminum silicate)
  B. Silica-Based
    1. Minerals (sand, quartz, diatomaceous earth)
    2. Synthetic

  C. Metals
  D. Boron filaments
  E. Glass
    1. Solid and hollow glass spheres
    2. Milled fibers
    3. Flakes
    4. Fibrous glass
  F. Metallic oxides
    1. Ground fillers (zinc oxide, titania, magnesia, alumina)
    2. Whiskers (alumina, magnesia, thoria, zirconia, beryllia)
  G. Calcium Carbonate
    1. Limestone
    2. Chalk
    3. Precipitated calcium carbonate
  H. Other Fillers
    1. Whiskers (nonoxides) (aluminum nitride, boron carbide, silicon carbide, silicon nitride, tungsten carbide, beryllium carbide)
    2. Barium sulfate
    3. Barium ferrite
    4. Potassium itanate

also been obtained by embedding carbon black in the surfaces of nylon or polyester filament reinforcements. The resistance of most polymers to ultraviolet radiation is also improved by the incorporation of carbon black.

Although glass spheres are classified as nonreinforcing fillers, the addition of 40 g of these spheres to 60 g of nylon 66 increases the flexural modulus, the compressive strength, and the melt index of the polymer. The tensile strength, the impact strength, the creep resistance, and the elongation of these composites are less than those of the unfilled nylon 66.

Zinc oxide (ZnO) is widely used as an active filler in rubber and as a weatherability improver in polyolefins and polyesters. Titanium dioxide (TiO_2) is widely used as a white pigment and as a weatherability improver in many polymers. Ground barites ($BaSO_4$) yield x-ray–opaque plastics with controlled densities. The addition of finely divided calcined alumina or silicon carbide produces abrasive composites. Zirconia, zirconium silicate, and iron oxide, which have specific gravities greater than 4.5, are used to produce plastics with controlled high densities.

Silica is used as naturally occurring and synthetic amorphous silica, as well as in the form of large crystalline particulates, such as sand and quartz. Diatomaceous earth, also called infusorial earth, fossil flour, and Fuller's earth, is a finely divided naturally occurring amorphous silica consisting of the skeletons of diatoms. Diatomaceous earth is used to prevent rolls of film from sticking together (antiblocking) and to increase the compressive strength of polyurethane (PU) foams.

Pyrogenic or fumed silica is a finely divided filler which is used as a thixotrope to increase the viscosity of liquid resins.

Sharp silica sand is used as a filler in resinous cement mortars and to provide an abrasive surface in polymers. Reactive silica ash, produced by burning rice hulls, and a lamellar filler, novaculite, which is obtained from the novaculite uplift in Arkansas, are also used as silica fillers in polymers.

Clay is used as a filler in compounding paper and rubber. Talc, a naturally occurring fibrouslike hydrated magnesium silicate, is used to improve the thermal resistance of polypropylene (PP). Since talc-filled PP is much more resistant to heat than unfilled PP, it is used in automotive accessories subject to high temperatures. Over 40 million tons of talc are used annually as a filler.

Conductive composites are obtained when powdered metal fillers or metal-plated fillers are added to resins. These composites have been used to produce forming tools for the aircraft industry. Powdered lead-filled polyolefin composites have been used as shields for neutron and gamma radiation, and metal-filled plastics have been used to prevent interference from stray electrons (EMI).

Among the naturally occurring filler materials are cellulosics, such as wood flour, alpha cellulose, shell flour, and starch, and proteinaceous fillers, such as soybean residues. Approximately 40,000 tons of cellulosic fillers are used annually by the U.S. polymer industry. Wood flour, which is produced by the attrition grinding of wood wastes, is used as a filler for phenolic resins, dark-colored urea resins, polyolefins, and polyvinyl chloride (PVC). Shell flour, which lacks the fibrous structure of wood flour, is made by grinding walnut and peanut shells. It is used as a replacement for wood flour.

Cellulose, which is more fibrous than wood flour, is used as a filler for urea and melamine plastics. Melamine dishware is a laminated structure consisting of molded resin-impregnated paper. Starch and soybean derivatives are biodegradable, and the rate of disintegration of resin composites containing these fillers may be controlled by the amount of these fillers present in polymers.

11.3 Reinforcements

Fillers with high aspect ratios (l/d), such as fibers, are classified as reinforcements. The strongest unidirectional reinforced composites are obtained when reinforcing fibers are continuous or at least longer than the critical length lc. The strength of these composites is reduced to 33% when the reinforcing fibers are randomly oriented, and there is an additional reduction in mechanical properties when the fiber length is less than lc.

The strongest composites are made from continuous filaments impregnated with resin before curing. These continuous filaments are wound around a mandrel in the filament-winding process and gathered together and forced through an orifice in the pultrusion-molding process.

The first continuous filaments were rayon, and these, as well as polyacrylonitrile (PAN) fibers, have been pyrolyzed to produce graphite fibers. High-modulus reinforcing filaments have also been produced by the deposition of boron atoms from boron trichloride vapors onto tungsten or graphite filaments.

Small single crystals, such as those of potassium titanate, are being used at an annual rate of over 10,000 tons for the reinforcement of nylon and other thermoplastics. These composites are replacing die-cast metals in many applications. Another microfiber, sodium hydroxycarbonate (Dawsonite), also improves the physical properties and flame resistance of many polymers. Many other single crystals, called whiskers, such as alumina, chromia, and boron carbide, have been used for making high-performance composites.

Polyester-resin-impregnated fibrous glass is used as a sheet-molding compound (SMC) and a bulk-molding compound (BMC). The latter is used like a molding powder, and the former is hot-pressed in the shape of the desired object, such as half a suitcase. Chopped fibrous glass roving may be impregnated with resin and sprayed, and glass mats may be impregnated with resin just prior to curing.

Graphite is an excellent but expensive reinforcement for plastics. Aramid (aromatic polyamide), polyester (polyethylene terephthalate; PET), and boron filaments are also used as reinforcements for polymers.

11.4 Coupling Agents

The high strength of polymer composites is dependent on a transfer of external stress from the resin matrix to the fiber or filler. The efficiency of this transfer is dependent on the strength of the interfacial bond between the resin and the fiber surface. The strength of this bond is improved by the addition of surface-active or coupling agents such as alkylsilanes, organotitanates, as well as high-molecular-weight carboxylic acids and esters, which are attracted by both the resin and the fiber surface.

The original coupling agents, which were called *promotors,* were used to ensure a good bond between rubber and the carbon black filler. These promotors increase the tensile strength, modulus, and the bound rubber (the insoluble mixture of filler and rubber) content of rubber. Although natural rubber is soluble in benzene, it becomes less soluble when carbon black or amorphous silica is added.

Many different proprietary alkylsilanes and organotitanates have been developed for use with specific composites. These coupling agents have two different functional groups; one is attracted to the resin and the other to the surface of the filler. For example, dialkyldimethoxysilanes are hydrolyzed to produce dialkyldihydroxysilanes *in situ.* It is believed that the hydroxyl groups bond with the filler surface and that the alkyl groups are attracted to the resin.

Filler Surface

$$H_3C-O \qquad O-CH_3 \qquad HO \qquad OH$$
$$\diagdown \qquad \diagup \qquad \xrightarrow{H_2O} \qquad \diagdown \qquad \diagup$$
$$Si \qquad\qquad\qquad Si$$
$$\diagup \qquad \diagdown \qquad\qquad \diagup \qquad \diagdown$$
$$R \qquad R \qquad\qquad R \qquad R$$

Resin

11.5 Antioxidants

Secondary aromatic amines, such as phenyl beta-naphthylamine, have been used as antioxidants in elastomers, but the preferred antioxidants for plastics have been 2,6-disubstituted and 2,4,6-trisubstituted phenols. These hindered phenols serve as chain transfer agents with the macroradicals which are produced by the degradation of polymers.

PP cannot be used out-of-doors unless a small amount of stabilizer (antioxidant) is present. It is known that the tertiary hydrogen atoms on every other carbon atom in the repeating units may be easily cleaved to form free radicals (R·). As shown following,

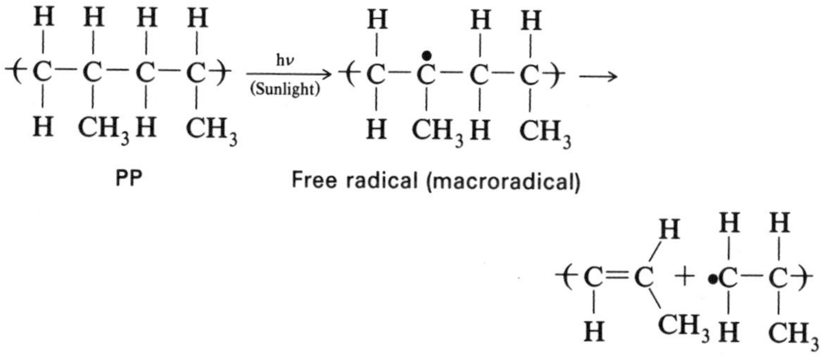

$$\begin{array}{c} \underset{\underset{\overset{|}{H}}{\overset{\overset{|}{H}}{}}}{C} - \underset{\underset{\overset{|}{CH_3}}{\overset{\overset{|}{H}}{}}}{C} - \underset{\underset{\overset{|}{H}}{\overset{\overset{|}{H}}{}}}{C} - \underset{\underset{\overset{|}{CH_3}}{\overset{\overset{|}{H}}{}}}{C} \xrightarrow[\text{(Sunlight)}]{h\nu} \end{array}$$

PP

Free radical (macroradical)

$$\text{Lower-molecular weight polymer} \qquad \text{Lower-molecular weight macroradical}$$

this carbon–hydrogen cleavage can lead to the breakage of the carbon–carbon bonds in the polymer chain in a process called chain or polymer degradation. This degradation process eventually causes the polymer to fail, break, and crumble.

The deterioration resulting from the formation of these free radicals is lessened when antioxidants, such as hindered phenols, alkyl phosphites $[(RO)_3P]$, or thioesters such as $ROOCCH_2CH_2SCH_2CH_2COOR$ are present:

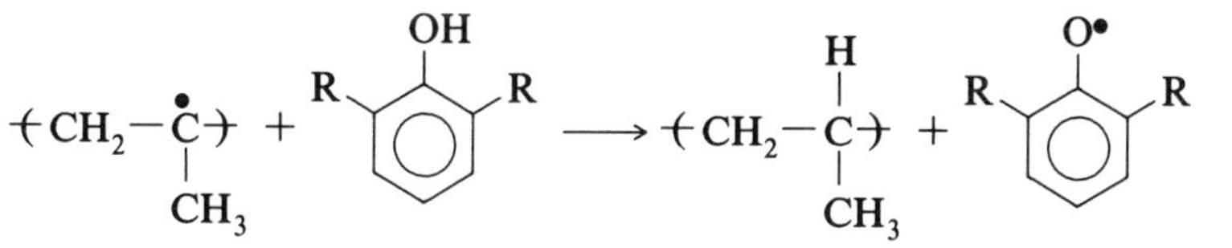

PP Macroradical   Hindered phenol   PP   Stable free radical

11.6 Ultraviolet Light Stabilizers

Although antioxidants trap the free radicals produced by the degradation of polymers, radiation stabilizers absorb radiation prior to molecular bond breakage.

The bond energy of most molecular (primary) bonds is on the order of 50 to 120 kcal per mole of bonds. Much of the sun's high-energy radiation is absorbed by the outer atmosphere, but some radiation in the 280 to 400 nm (ultraviolet) range gets through to the earth's surface. The energy of this radiation is 100 to 72 kcal and is sufficiently strong enough to cleave covalent bonds and cause yellowing and embrittlement of organic polymers. This degradation of polymers may be minimized when compounds which absorb this high ultraviolet energy are present.

Some stabilizers, such as 2-hydroxybenzophenone, produce cyclic compounds (chelates) which absorb the ultraviolet energy and release it at a lower, less destructive energy level. A chelate is a 5- or 6-membered ring-like specie which may be formed by intramolecular attractions of a hydrogen atom to an oxygen atom in the same compound. Thus when the hydroxyl group of 2-hydroxybenzophenone is in the ortho position, it forms a chelate with the carbonyl oxygen, as shown in the following formula:

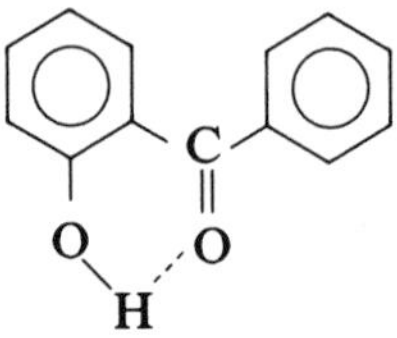

11.7 Flame Retardants

Although halogen-containing polymers, such as PVC and PTFE, and phosphorus-containing phosphonate and phosphate polymers

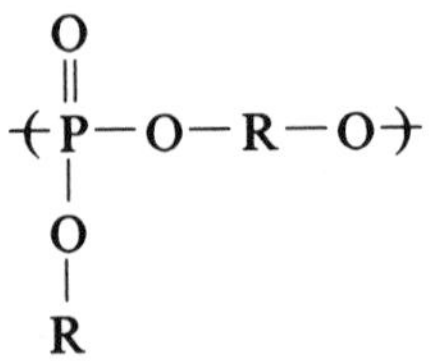

Polyphosphate Esters 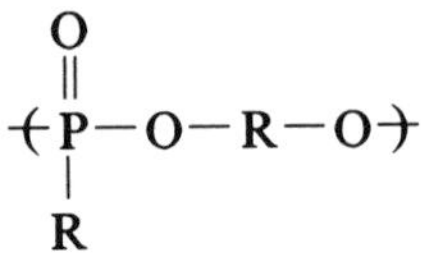   Polyphosphonate Esters

are considered flame-retardant in air, they will burn under some conditions. Most general-purpose polymers, such as polyethylene, nylon, and PS, are combustible; when they are used for construction, clothing, and furniture, it is essential that their combustibility be greatly decreased by the addition of flame retardants.

Fuel, oxygen, and a high temperature are essential for the combustion process. Polyfluorocarbons, polyphosphazenes, and some polymer composites have flame-retardant properties and are not good fuels. Some fillers, such as ATH, release water when heated and thus reduce the temperature of the combustion reaction. Compounds which release carbon dioxide, such as sodium carbonate, shield the reactants from oxygen.

Char also shields the reactants from oxygen and in addition retards the outward diffusion of volatile combustible products. Aromatic polymers tend to char, and some phosphorus and boron compounds tend to catalyze char formation.

Synergistic flame retardants, such as a mixture of antimony trioxide and an organic halogen compound, are much more effective than single flame retardants. Thus although a flame retardant polyester containing 11.5% tetrabromophthalic anhydride burns without charring at high temperatures, charring but no burning at high temperatures is noted when 5% antimony trioxide is added.

Since combustion is subject to many variables, tests for flame retardancy may not predict flame resistance under unusual conditions. Thus a disclaimer stating that flame-retardant tests do not predict performance in an actual fire must accompany all flame-retardant polymeric materials. Flame retardants, like many other organic compounds, may be toxic, or they may produce toxic gases when burned. Thus care must be exercised when using fabrics or other polymers treated with flame retardants.

11.8 Plasticizers

Since cellulose nitrate is intractable, in 1870 John W. Hyatt added camphor as a plasticizer to flexibilize this plastic. Some 60 years later, Waldo Semon used tricresyl phosphate as a plasticizer for PVC. Dialkyl phthalates, such as dioctyl phthalate (DOP) and other alkyl phthalates which replaced the more toxic tricresyl phosphate, are now used as plasticizers primarily for PVC at an annual rate of 1 million tons.

Plasticizers are compounds which increase the flexibility and processibility of polymers. It has been postulated that the added plasticizer reduces the intermolecular forces in PVC and increases the free volume. Effective plasticizers, like effective solvents, have solubility parameters within 1.8 H (Hildebrand units) of that of the polymer.

Plasticizers are usually moderately large, nonvolatile molecules, such as DOP:

$$\text{(benzene ring)} \quad \overset{\text{COO} \!+\! CH_2 \!\cdot\!\!\cancel{)}_7 CH_3}{\underset{\text{COO} \!+\! CH_2 \!\cdot\!\!\cancel{)}_7 CH_3}{}}$$

Most plasticizers solubilize polymer units and improve segmental movement but do not promote wholesale chain movements.

The development of plasticizers has been plagued with toxicity problems. Thus the use of highly toxic polychlorinated biphenyls (PCBs) has been discontinued. Blood stored in plasticized PVC blood bags and tubing may extract phthalic acid esters, such as DOP. These aromatic esters are also distilled slowly from PVC upholstery in closed automobiles in hot weather. These problems have been solved by using oligomeric polyesters as nonfugitive plasticizers instead of DOP.

Many copolymers are said to be internally plasticized because of the flexibilization brought about by the presence of a second repeating unit in the polymer chain. In contrast, DOP and other liquid plasticizers are said to be external plasticizers. The presence of bulky pendant groups on the polymer increases segmental motion, and the flexibility of the polymer increases as the size of the pendant group increases. However, linear pendant groups with more than 10 carbon atoms reduce flexibility because of side chain crystallization.

Plasticizer containment still remains a major problem, particularly for periods of extended use. For instance, most plastic floor tiles become brittle with extended use, mainly due to the leaching out of the plasticizer. This problem has been solved, to some extent, through many routes, including surface treatment of polymer products and the use of branched polymers which are more flexible than linear polymers.

The annual worldwide production of plasticizers is over 3 million tons, and the U.S. production is in excess of 1 million tons. In fact, plasticizers are major components of a number of polymer-containing products. For instance, the liner in automobile safety glass is composed mainly of polyvinyl butyral containing about 30% plasticizer.

11.9   Heat Stabilizers

In addition to the free-radical chain degradation described for polyolefins, another type of degradation (dehydrohalogenation) also occurs with chlorine-containing polymers, such as PVC. As shown by the following equation,

$$
\begin{array}{ccccccc}
\text{H} & \text{H} & \text{H} & \text{H} \\
| & | & | & | \\
\text{+C} & \text{–C} & \text{–C} & \text{–C+} \\
| & | & | & | \\
\text{H} & \text{Cl} & \text{H} & \text{Cl}
\end{array}
\longrightarrow
\begin{array}{ccccccc}
\text{H} & \text{H} & \text{H} & \text{H} \\
| & | & | & | \\
\text{+C} & \text{=C} & \text{–C} & \text{–C+} \\
 & & | & | \\
 & & \text{H} & \text{Cl}
\end{array}
+ \text{ HCl}
$$

PVC   Dehydrohalogenated PVC   Hydrogen chloride

PVC, when heated, may lose hydrogen chloride and form an alkene double bond, which is susceptible to further degradation, and eventually produces a variety of degradation products.

The tendency for PVC to degrade by the unzipping type of dehydrochlorination is reduced by the presence of heat stabilizers, such as barium and cadmium salts of high-molecular-weight carboxylic acids and epoxidized unsaturated aliphatic esters. The general reaction for cadmium salts is as follows:

$$
\begin{array}{c}
\text{H} \\
| \\
\text{C} \\
| \\
\text{Cl}
\end{array}
+ \;
\text{Cd(O}-\underset{\underset{\text{O}}{\|}}{\text{C}}-\text{R)}_2
\longrightarrow
\begin{array}{c}
\text{H} \\
| \\
\text{C} \\
| \\
\text{O}
\end{array}_{\searrow \underset{\text{O}}{\nearrow}\text{C}-\text{R}}
+ \; \text{CdCl}_2
$$

It should be noted that many types of stabilizers are employed, and each polymer system is unique and requires its own particular heat-stabilizer additives.

11.10 Impact Modifiers

Several flexible polymers, such as natural rubber (NR); synthetic rubber (SR); polyalkyl acrylates; copolymers of acrylonitrile, butadiene, and styrene, (ABS); and polyvinyl alkyl ethers, have been used to improve the impact resistance of PS and PVC. PS and copolymers of ethylene and propylene have been used to increase the ductility of polyphenylene oxide (PPO) and nylon 66, respectively. The mechanical properties of several other engineering plastics have been improved by blending them with thermoplastics.

11.11 Other Additives

Catalysts, colorants, foaming agents, biocides, lubricants, and antistats are also used as additives for polymers. Although foaming agents reduce the specific gravity, the other cited additives, when used in moderate amounts, have little effect on the physical or thermal properties of the polymers.

11.12   References

B. D. Agawal and L. J. Broutman, *Analysis and Performance of Fiber composites,* Wiley-Interscience, New York (1980).

V. Bhatnagar, *Advances in Fire Retardants,* Technomic, Westport, Conn. (1973).

P. F. Bruins, *Plasticizer Technology,* Vol. 1, Reinhold, New York (1974).

P. F. Bruins, *Polyblends and Composites,* Wiley-Interscience, New York (1970).

D. N. Buttrey, *Plasticizers,* 2nd ed., Franklin Publishing, Palisades, N.J. (1960).

H. T. Corten, ed., *Composite Materials: Testing and Design,* American Society for Testing and Materials (No. 497), Philadelphia (1972).

R. D. Deanin, *Thermal Stabilizers for PVC,* Plastics Institute of America Audio Course, Hoboken, N.J. (1979).

J. Delmonte, *Metal-Filled Plastics,* Reinhold, New York (1961).

A. K. Doolittle, *The Technology of Solvents and Plasticizers,* Wiley, New York (1954).

R. F. Gould, ed., *Plasticization and Plasticizer Processes,* American Chemical Society, Washington, D.C. (1965).

R. F. Gould, ed., *Stabilization of Polymers and Stabilizer Processes,* American Chemical Society, Washington, D.C. (1968).

C. J. Hilado, *Flammability of Fabrics,* Technomic, Westport, Conn. (1974).

H. S. Katz and J. V. Milewski, *Handbook of Fillers and Reinforcements for Plastics,* Van Nostrand-Reinhold, New York (1978).

G. Kraus, *Reinforcement of Elastomers,* Wiley-Interscience, New York (1965).

W. C. Kuryla and A. J. Papa, *Flame Retardancy of Polymeric Materials,* Dekker, New York (1973).

M. Langley, *Carbon Fibers in Engineering,* McGraw-Hill, New York (1973).

M. Lewis, S. M. Atlas, and E. M. Pearce, *Flame Retardant Polymeric Materials,* Plenum Press, New York (1975).

G. Lubin, *Handbook of Fiberglass and Advanced Plastics Composites,* Van Nostrand-Reinhold, New York (1969).

G. Lubin, *Handbook of Composites,* Van Nostrand-Reinhold, New York (1982).

J. H. Manson and L. H. Sperling, *Polymer Blends and Composites,* Van Nostrand-Reinhold, New York (1969).

I. Mellan, *Industrial Plasticizers,* Macmillan, New York (1969).

L. E. Nielson, *Mechanical Properties of Polymers and Composites,* Dekker, New York (1974).

S. S. Oleesky and J. J. Mohr, *Handbook of Reinforced Plastics,* Reinhold, New York (1973).

N. J. Parrott, *Fiber Reinforced Materials Technology,* Van Nostrand-Reinhold, New York (1973).

T. C. Patton, *Pigment Handbook,* Wiley-Interscience, New York (1974).

R. B. Seymour, ed., *Additives for Plastics,* Vols. 1 and 2, Academic Press, New York (1978).

R. B. Seymour, *Fillers and Other Additives for Plastics,* Audio Course Plastics Institute of America, Hoboken, N.J. (1981).

12 | Properties of Polyolefins

12.1 Introduction

The principal polyolefins are low-density polyethylene (LDPE), high-density polyethylene (HDPE), linear low-density polyethylene (LLDPE), polypropylene (PP), polyisobutylene (PIB), poly-1-butene (PB), copolymers of ethylene and propylene (EP), and proprietary copolymers of ethylene and alpha olefins. Since all these polymers are aliphatic hydrocarbons, the amorphous polymers are soluble in aliphatic hydrocarbon solvents with similar solubility parameters. Like other alkanes, they are resistant to attack by most ionic and most polar chemicals; their usual reactions are limited to combustion, chemical oxidation, chlorination, nitration, and free-radical reactions.

12.2 High-Density Polyethylene

Linear polyethylene (HDPE), which has a repeating unit of $-(CH_2-CH_2)-$, is a typical thermoplastic with a glass transition temperature T_g below room temperature. Its mechanical properties permit its use as both a plastic and a fiber. From a structural viewpoint, this polymeric hydrocarbon is the least complex of all polymers.

Because of its structural regularity and lack of pendant groups, HDPE exists in a zigzag planar structure and is a highly crystalline polymer. HDPE has fewer than 7 ethyl groups per 1000 carbon atoms. It has a relatively high specific gravity (0.96). The specific gravity of a perfect single crystal of HDPE is about 1.0.

The index of refraction of amorphous HDPE is 1.49, but that of highly crystalline HDPE is greater than 1.52. Commercial HDPE is soluble in hot xylene, but this highly crystalline polymer is insoluble in most solvents at room temperature.

Although low-molecular-weight HDPE is slightly brittle, the high-molecular-weight polymer is more ductile, and extremely high molecular weight HDPE has outstanding resistance to impact. Since the intermolecular forces present in these nonpolar molecules are weak dispersion (London) forces, the modulus and the tensile strength of HDPE are relatively low.

HDPE has a melting point T_m of at least 125°C, and because of its weak intermolecular forces and high flexibility, it tends to creep when subjected to high compressive forces over long periods of time. The compressive strength of a typical HDPE plastic is 20,000 kPa. The Izod impact strength is about 30 cm · N per centimeter of notch. HDPE has a tensile strength of 27,000 kPa and an elongation of 30%.

HDPE has a Rockwell hardness of D40 and a coefficient of linear expansion of 12×10^{-5} cm/cm · °C, but this value may be reduced by the addition of fillers. Because of the absence of polar groups, HDPE is an excellent electric insulator.

HDPE has a dielectric constant of 2.3 at 60 Hz and higher frequencies. Its power factor is less than 0.0005 at 60 Hz.

Since HDPE is a linear hydrocarbon polymer and, like linear alkanes, sputters when ignited, it burns readily unless admixed with alumina trihydrate (ATH) or other flame retardants. It has a solubility parameter of 7.9 H and low water absorption (0.01%).

Infrared spectroscopic studies show that HDPE produced by Ziegler-Natta catalysis has about 10 to 14 ethyl groups per 1000 repeating units. However, HDPE produced by the use of the Phillips catalyst has less branching and fewer ethyl groups. Most of the commercial HDPE used in the United States is produced by use of the Phillips catalyst.

HDPE crystallizes readily and is not easily permeated by gases and vapors. The carbon dioxide permeation in HDPE is three times as fast as the oxygen permeation, which in turn is four times as fast as the nitrogen permeation. HDPE has fair resistance to environmental stress cracking because of its highly crystalline nature and nonpolar properties. This resistance is increased as the number average molecular weight of the polymer increases.

HDPE has a moderately high heat deflection temperature (85°C) and a moderately high T_m (130°C). HDPE has good resistance to alkalis, nonoxidizing acids, and aqueous salt solutions. It is attacked by strong oxidizing acids such as nitric acid.

As shown in Table 12.1, the impact resistance and tensile strength of ultrahigh-molecular-weight polyethylene (UHMWPE) are superior to those of HDPE, but UHMWPE is more difficult to process. UHMWPE is also slightly harder than HDPE.

The addition of fillers and reinforcing agents, such as glass fibers, increases the heat deflection temperature and lowers the coefficient of linear expansion for HDPE. As shown in Table 12.1, HDPE containing 30% glass fibers has a heat deflection temperature of 120°C and a coefficient of linear expansion of 5.0 $\times$ 10^{-5} cm/cm · °C. The filled polymer also has a higher specific gravity, is harder, and has better strength properties than unfilled HDPE.

12.3 Low-Density Polyethylene

As shown in Figure 12.1, LDPE is a highly branched polymer. Infrared spectroscopic studies indicate that there are about 60 ethyl and butyl groups for a molecule of LDPE with a DP of 1000.

The reduction in symmetry resulting from the higher degree of branching impedes crystallization. As a result of this branching, LDPE is more flexible, more transparent (less crystalline), and more ductile than HDPE.

Table 12.1 Polyethylenes[a]

Thermal and physical properties of plastics (typical values)	LDPE	HDPE	UHMWPE	30% glass-filled HDPE
Heat deflection temperature @ 1820 kPa (°C)	40	50	85	120
Maximum resistance to continuous heat (°C)	40	80	80	110
Coefficient of linear expansion cm/cm · °C $\times$ 10^{-5}	10.0	12.0	12.0	5.0
Compressive strength (kPa)	—	20,000	—	43,000
Flexural strength (kPa)	—	—	—	75,800
Impact strength (Izod: cm · N/cm of notch)	No break	30	No break	53
Tensile strength (kPa)	5,515	27,580	38,000	62,000
Elongation (%)	100	30	400	1.5
Hardness: Rockwell	D40	D40	R50	R75
Specific gravity	0.91	0.96	0.94	1.3

[a]It must be noted that the values cited in the tables of data and within the text for polymeric materials are typically average values with the actual measured value obtained for a material dependent on the specific origin of the material (i.e., conditions of synthesis and pretreatment) and on the specific test procedure employed.

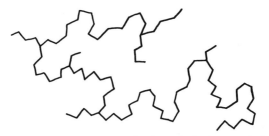

Figure 12.1   A portion of a hypothetical LDPE chain

Relatively clear thin films of LDPE may be obtained by the addition of a nucleating agent, such as sodium benzoate, and/or by quenching hot films.

LDPE has a relatively low tensile strength (5,515 kPa) and a relatively high elongation (at least 100%) at the yield point. Its Rockwell hardness of D40 is the same as that of HDPE, but its specific gravity is lower (0.91).

LDPE has a lower coefficient of expansion (10×10^{-5} cm/cm · °C) and a lower heat deflection temperature (40°C) than HDPE. LDPE does not break when subjected to the Izod impact test. The T_g of LDPE is about -120°C.

Like HPDE, the lower-density polyethylene has very good electric properties. The dielectric constant of LDPE is 2.2. Since it is a hydrocarbon, LDPE burns readily, but because of its branched structure, it sputters less when burning than HDPE.

LDPE has a solubility parameter of 7.9 H, but because it has a degree of crystallinity, it is resistant to nonpolar solvents at ordinary temperatures. Like HDPE, because of its nonpolarity it has a low water absorption value (less than 0.01%).

Because of its branched structure and lower degree of crystallinity, LDPE is about four times more permeable to carbon dioxide, oxygen, and nitrogen than HDPE.

LDPE may stress crack when exposed to nonpolar solvents or surface-active agents. Like HDPE, this nonlinear polyethylene is resistant to nonoxidizing acids, alkalis, and salts but is attacked by oxidizing corrosives, such as nitric acid.

12.4   Copolymers of Ethylene

Ionomers are random copolymers of ethylene and methacrylic acid in which the formula for some of the repeating units is as follows:

$$\begin{array}{c} CH_3 \\ | \\ +CH_2-CH_2-C-CH_2+ \\ | \\ C=O \\ | \\ OH \end{array}$$

These polar copolymers are more transparent and, as a result of the presence of the polar acrylic acid moiety, have better adhesion to metallic surfaces than LDPE. The commercial copolymer, which is usually 50% neutralized to salts of sodium or zinc, is stiff and strong because of "ionic crosslinking" at room temperature, but it is readily processible when heated.

Ionomers are superior to LDPE in resistance to environmental stress cracking, but, because ionomers are more polar, they are not as useful as electric insulators. Blow-molded containers made from ionomers, unlike those made from LDPE or HDPE, do not store static electricity and hence do not attract dust.

Random copolymers of ethylene and vinyl acetate have the following repeating units:

$$\begin{array}{c} +CH_2-CH_2+_x+CH_2-CH+_y \\ | \\ O-C-CH_3 \\ || \\ O \end{array}$$

They are more transparent than HDPE and are less transparent, less crystalline and more rubbery than LDPE. They may be used as plastic melts.

These copolymers are twice as permeable to carbon dioxide, oxygen, and nitrogen as LDPE. They have lower tensile strength and slightly higher water absorption than LDPE.

Random copolymers of ethylene and 1-butene or propylene are more flexible and more resistant to environmental stress cracking than LDPE. Some of the newer commercial LLDPE plastics are copolymers of ethylene and alpha alkenes, such as 1-hexene.

12.5 Crosslinked Polyethylene

Polyethylene may be crosslinked by electron beams or by free-radical initiators. These polymers have *elastic memory*, i.e., stretched crosslinked films or tubing shrinks to the original dimensions when heated.

12.6   Polypropylene

The repeating unit in PP is as follows:

$$\begin{array}{c} CH_3 \\ | \\ +CH_2-CH+ \end{array}$$

Atactic (*at*) PP lacks stereoregularity and is a soft, transparent, viscous liquid. This polymer has little use except as a softener for other polymers. Unlike the more crystalline isotactic *it*-PP, *at*-PP is soluble in hexane.

it-PP, in which all the methyl pendant groups have a similar arrangement around the chain carbon atoms, as shown by the following abbreviated formula,

$$\begin{array}{c} H_3C \quad H \; H_3C \quad H \; H_3C \quad H \; H_3C \quad H \; H_3C \quad H \\ \backslash \quad / \quad \backslash \quad / \quad \backslash \quad / \quad \backslash \quad / \quad \backslash \quad / \\ \sim\!\!+CH_2-C+\;CH_2-C-CH_2-C-CH_2-C-CH_2-C-\sim \end{array}$$

is a crystalline polymer in which the chains are present in helical conformations, with three repeating units per turn. Because of the added volume which is left between the pendant methyl groups, this opaque polymer has a low specific gravity (0.90) as cited in Table 12.2. Its index of refraction is 1.49.

Because of the orderly arrangement of pendant methyl groups, which allows the units to fit well together, rotation of the chain is restricted. *it*-PP has a relatively high T_g (-18 °C) and a relatively high T_m (176 °C).

Commercial PP (*it*-PP) has a higher tensile strength (34,500 kPa) than HDPE but a lower Izod impact strength (27 cm·N per centimeter of notch).

Table 12.2   Polypropylene

Thermal and physical properties of plastics (typical values)	Unfilled PP	40%-Talc-filled PP
Heat deflection temperature @ 1820 kPa (°C)	55	100
Maximum resistance to continuous heat (°C)	100	120
Coefficient of linear expansion cm/cm · °C $\times$ 10^{-5}	9.0	6.5
Compressive strength (kPa)	44,800	54,000
Flexural strength (kPa)	48,000	58,600
Impact strength (Izod: cm · N/cm of notch)	27	27
Tensile Strength (kPa)	34,500	30,000
Elongation (%)	100	5
Hardness: Rockwell	R80	R95
Specific gravity	0.90	1.22

High-molecular-weight PP is also more ductile than HDPE, and its coefficient of expansion is slightly lower (9×10^{-5} cm/cm · °C). Its heat deflection temperature (55 °C) is higher than that of HDPE.

it-PP filled with 40% talc has a lower coefficient of linear expansion (6.5×10^{-5} cm/cm · °C), higher heat deflection temperature (100 °C), and higher specific gravity (1.22) than the unfilled polymer.

The permeability of it-PP to gases is similar to that of HDPE. It has very good electric properties and low moisture absorption. PP burns readily without sputtering.

When heated, PP dissolves in nonpolar solvents with solubility parameters in the range of 6.5 to 9.5 H, but it is resistant to polar solvents, such as ethanol. PP is more resistant to environmental stress cracking than polyethylene.

PP is resistant to nonoxidizing acids, alkalis, and salts but is attacked by oxidizing reagents, such as nitric acid. In the absence of stabilizers, it is readily degraded when exposed to sunlight. Unlike polyethylene, which crosslinks when exposed to radiation, PP degrades when exposed to high-energy electron beam radiation.

12.7 Copolymers of Propylene

The random copolymer of propylene and ethylene (EP) lacks the good symmetry of it-PP and is a flexible elastomer. Since this copolymer is used as an elastomer, it is customary to add a small amount of a diene, such as ethylidene norbornene, to the monomer reactants before polymerization to allow subsequent cross-linking or curing.

This three-component copolymer (terpolymer) (EPDM) contains about 15 double bonds per 1000 carbon atoms in the polymer. This provides enough reactive sites for low-density cross-linking, but unlike *Hevea* rubber (polyisoprene), which contains about 200 double bonds per 1000 carbon atoms, EPDM is resistant to ozone. EPDM is used for the sidewalls of tires and for rubber hose but develops too much heat during flexing to be used for heavy-duty tire treads.

Copolymers of propylene and a small amount of ethylene (3%) are called *polyallomers*. These copolymers are not as flexible as EPDM, but they are more ductile than it-PP.

Copolymers of propylene are used as additives for improving the impact resistance of stiff polymers, such as nylon 66. Some of the new readily processible commercial PPs are copolymers of propylene and alpha alkenes.

12.8   Polyisobutylene

Polyisobutylene (PIB), which has the following repeating unit,

$$+CH_2-\underset{\underset{CH_3}{|}}{\overset{\overset{CH_3}{|}}{C}}+$$

is an amorphous elastomer with a low T_g ($\cong -70$ °C). Because of its tendency to cold flow, its use is limited to adhesives, chewing gum, caulking compositions, and coatings, and as flexibilizing additives for rigid polymers and as oil additives.

The copolymer of isobutylene with a few percent isoprene (butyl rubber) can be cured to produce an ozone-resistant elastomer with low permeability to oxygen and nitrogen. Butyl rubber has a T_g of -70 °C, a refractive index of 1.5081, and a coefficient of linear expansion of 5.7×10^{-5} cm/cm · °C. Chloro and bromo butyl rubber are more resistant to the permeation of oxygen and nitrogen than butyl rubber.

12.9   Poly-1-butene

PB, which has the following repeating unit,

$$+CH_2-\underset{\underset{\underset{CH_3}{|}}{\overset{CH_2}{|}}}{\overset{\overset{CH_2}{|}}{CH}}+$$

is obtained directly by the Ziegler-Natta–catalyzed polymerization of 1-butene. Hydrogenation of it-poly-1,2-butadiene yields it-poly-1-butene. These polymers are crystalline because of the presence of ethyl pendant groups which restrict rotation of the repeating units. it-PB exists as a loosely packed helix, but its rate of crystallization is much lower than that of PP.

PB has a specific gravity of 0.91 and a T_m of 126 °C. Like PP, it is readily oxidized because of the tertiary carbon atoms present. Since its solubility parameter is about 8.0 H, it is not soluble in polar solvents nor in low-molecular-weight alkanes such as pentane, but is soluble in hot alkanes with slightly higher molecular weights. PB is not attacked by nonoxidizing acids, alkalis, or salts.

12.10   Poly-4-methylpentene-1

Poly-4-methylpentene-1 (TPX) is a helically coiled crystalline polymer having the following repeating unit:

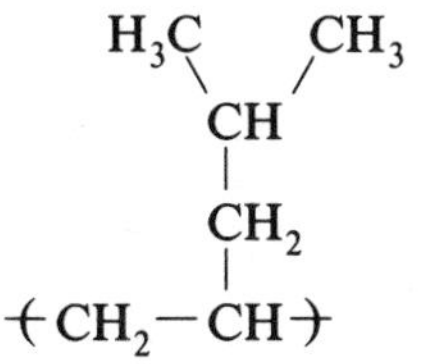

$$H_3C \quad CH_3$$
$$CH$$
$$|$$
$$CH_2$$
$$|$$
$$+CH_2-CH+$$

As shown in Table 12.3, TPX has an unusually low specific gravity (0.83) and a heat deflection temperature of 80 °C. Since the specific gravities of the amorphous and the crystalline polymer are similar, the polymer is transparent. Because of the loose packing resulting from the large regularly spaced pendant groups, TPX is quite permeable to many gases and water vapor. The commercial product is a copolymer with a small amount of an alpha alkene.

TPX has a coefficient of linear expansion of 11.7×10^{-5} cm/cm · °C, which is similar to that of water, and hence it is useful for calibrating containers such as laboratory graduated cylinders. TPX is resistant to nonoxidizing acids, alkalis, and salts but is not resistant to oxidizing acids, such as nitric acid. Because of the presence of tertiary carbon atoms, TPX, like PP, is readily oxidized and cannot be used outdoors in the absence of antioxidants.

Table 12.3   Poly-4-methylpentene-1 (TPX)

Thermal and physical properties (typical values)	
Heat deflection temperature @ 1820 kPa (°C)	80
Maximum resistance to continuous heat (°C)	40
Coefficient of linear expansion cm/cm · °C $\times 10^{-5}$	11.7
Compressive strength (kPa)	38,000
Flexural strength (kPa)	34,000
Impact strength (Izod: cm · N/cm of notch)	27
Tensile strength (kPa)	24,000
Elongation (%)	15
Hardness: Rockwell	L70
Specific gravity	0.83

12.11   Polyalkadienes

Cis-Poly-1,4-butadiene, which has the following repeating unit:

$$+CH_2-CH=CH-CH_2+$$

is an amorphous, flexible elastomer with a low degree of crystallinity and a low T_g ($-95\,°C$). As in the case of polyethylene, the intermolecular attractions are limited to weak dispersion forces. The linear low-molecular-weight polymer tends to cold flow, but because of entanglement in the higher-molecular-weight polymers, there is less tendency for the high-molecular-weight materials to cold flow.

Cured polymers of butadiene with low cross-link density do not tend to cold flow and are useful elastomers. These vulcanized elastomers crystallize when stretched, but when the stress is removed, the restoring force is largely entropy and most of the crystals melt and the chains return to the random conformation.The tensile strength is increased dramatically when large amounts of carbon black or amorphous silica are added.

Because of the absence of the methyl pendant group, *cis*-1,4-polybutadiene is more resistant to abrasion and develops less heat buildup as a result of cyclic flexing than *Hevea* rubber (polyisoprene; NR). Tires made from polybutadiene do not grip the road as well as those made from *Hevea* elastomers.

The products with low crosslink density are elastomers, but infusible, hard plastics are obtained when these polymers are crosslinked with large amounts of sulfur (40%).

Cis-poly-1,4-isoprene has a specific gravity of 0.91, a coefficient of linear expansion of 67×10^{-5} cm/cm · °C, and a refractive index of 1.5191. This and other elastomers retain their characteristic mobility at temperatures above the T_g. They are brittle at temperatures below T_g.

Like LDPE, polybutadienes are resistant to most nonoxidizing acids, alkalis, and salts. However, because they are unsaturated, the polyalkadienes are attacked by hydrochloric, hydrobromic, and hydrofluoric acids, as well as by hydrogen and chlorine. The reaction products, which are thermoplastic, have been used as commercial nonelastomeric plastics. NR and other diene elastomers are also attacked by peroxides and ozone. In the absence of antioxidants and carbon black filler, these unsaturated elastomers are degraded in the sunlight.

Because of the vinyl pendant group, *it*-poly-1,2-butadiene, which has the following repeating unit,

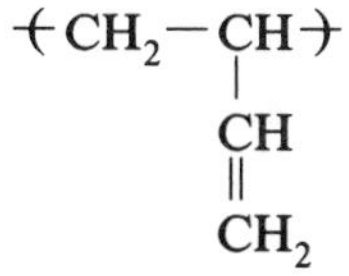

$$+CH_2-\underset{\underset{\overset{\displaystyle CH}{\displaystyle \|}}{\displaystyle CH}}{\displaystyle CH}+$$

has a relatively low specific gravity. It is more permeable to oxygen and nitrogen than *cis*-poly-1,4-butadiene. In the absence of stabilizers and in the presence of ultraviolet light, this transparent, highly amorphous polymer crosslinks to produce a brittle network polymer. It has a T_g of -4 °C.

Both the naturally occurring *Hevea brasiliensis* (NR) and synthetic *cis*-poly-1,4-isoprene have the following repeating unit:

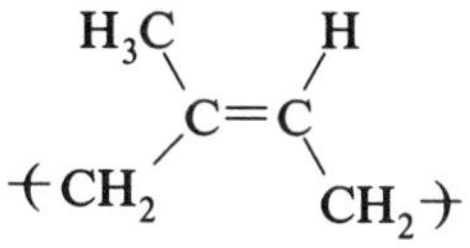

$$\underset{+CH_2}{\overset{H_3C}{\diagdown}}C=C\underset{CH_2+}{\overset{H}{\diagup}}$$

These highly amorphous elastomers have relatively low T_g values (-73 °C) and tend to crystallize when stretched. The cold flow of these thermoplastic polymers is reduced when they are crosslinked (vulcanized) with a small amount (2%) of sulfur. Since these polymers of isoprene have a solubility parameter of 8.0 H, they are resistant to polar solvents but are soluble in many aliphatic and aromatic hydrocarbon solvents. The cross-linked derivatives swell but do not dissolve in these solvents.

The repeating unit in *trans*-poly-1,4-isoprene is as follows:

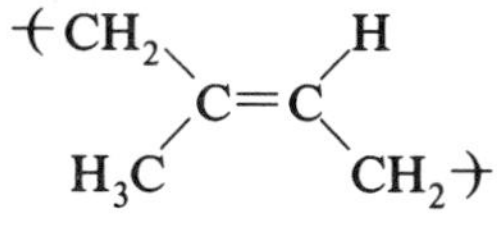

$$\underset{H_3C}{\overset{+CH_2}{\diagdown}}C=C\underset{CH_2+}{\overset{H}{\diagup}}$$

This polymer is a hard plastic that occurs naturally as gutta-percha or balata. Since the trans isomer packs better than the cis isomer, it has a higher specific gravity, a higher degree of crystallinity, and a higher melting point (67 °C) than the cis isomer of polyisoprene. The chemical and the solvent resistance of the trans polymer are similar to those of the cis polymer.

The random copolymer of butadiene (72%) and styrene (28%) (SBR) includes units of the following structure:

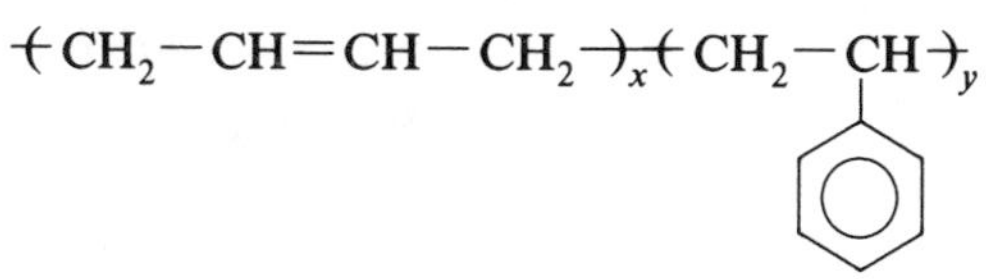

$$+CH_2-CH=CH-CH_2+_x+CH_2-CH+_y$$

SBR is the most widely used synthetic elastomer. It is an amorphous random copolymer consisting of a mixture of 1,2,; cis; and trans isomers. Cold SBR produced at -20 °C consists of 17% 1,2,; 6% cis; and 77% trans isomers of polybutadiene. This commercial product has a T_g of -60 °C, an index of refraction of 1.5345, and a coefficient of linear expansion of 66×10^{-5} cm/cm $\cdot$ °C. Because of the high percentage of the trans isomer, it is less flexible and has a higher heat buildup, when flexed, than *Hevea* rubber. Although carbon black-filled or amorphous silica-filled SBR has useful physical and mechanical properties, the SBR gum rubber is inferior to *Hevea* rubber.

The chemical and the solvent resistance of SBR are similar to those cited for *Hevea* rubber. SBR does not crystallize when stretched. However, its abrasion resistance is superior to that of *Hevea* rubber.

The copolymers of butadiene (55 to 82%) and acrylonitrile (45 to 18%) (Hycar) are oil- and heat-resistant elastomers which contain the following random repeating units:

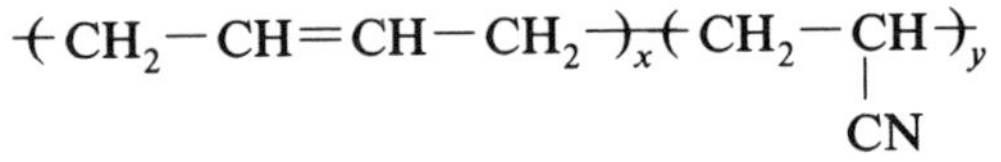

$$+CH_2-CH=CH-CH_2\xrightarrow{}_x(CH_2-\underset{\underset{CN}{|}}{CH})_y$$

The ABA block copolymer of styrene (12.5%)–butadiene (75%)–styrene (12.5%) (Kraton), is a thermoplastic elastomer (TPE) with the multiple repeating units shown below:

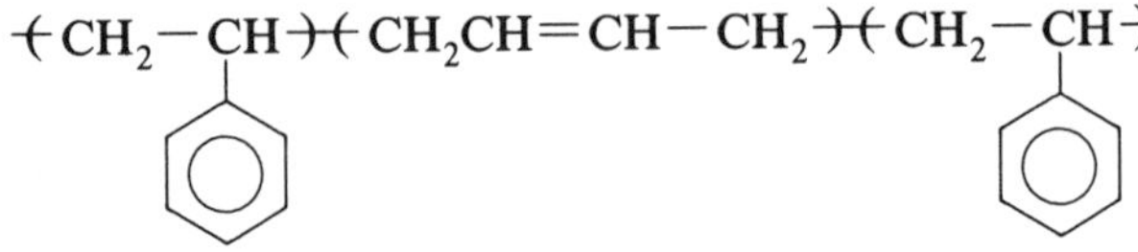

$$+CH_2-CH)+(CH_2CH=CH-CH_2)+(CH_2-CH)+$$

This ABA block copolymer consists of stiff polystyrene (PS) and resilient polybutadiene blocks. The domains of these TPEs have characteristic T_g values of 100 and -80 °C, respectively. The polybutadiene blocks retain their flexibility at low temperatures, and the polystyrene blocks lose their stiffness when the polymer is heated above 110 °C. A related thermoplastic is a transparent AB block copolymer of styrene and butadiene (K-resin).

These ABA copolymers have an index of refraction of 1.5 and water absorption of about 0.2%. Unless hydrogenated to saturated block copolymers, these unsaturated unstabilized plastics are degraded in sunlight. The polybutadiene domains are attacked by aliphatic hydrocarbon solvents, such as hexane, and the polystyrene domains are attacked by aromatic hydrocarbon solvents, such as benzene. These block copolymers are unaffected by polar solvents or nonoxidizing acids, alkalis, or salts. Hydrogenated ABA block copolymers are available commercially; they are actually block copolymers of styrene-1-butene and styrene.

12.12 References

S. Aggarwal, ed., *Block Copolymers*, Plenum Press, New York (1970).

D. Allport and W. H. James, *Block Copolymers*, Halsted, New York (1973).

H. J. Cantow, ed., *Polymer Chemistry*, Springer-Verlag, New York (1979).

R. J. Ceresa, ed., *Block and Graft Copolymerization*, (two volumes), Wiley, New York (1976).

H. D. Frank, *Polypropylene*, Gordon and Breach, New York (1968).

T. O. Kresser, *Polyolefin Plastics*, Van Nostrand-Reinhold, New York (1969).

G. Mathews, *Vinyl and Allied Polymers*, Vol. 2, Butterworth, London (1972).

A. Noshay and J. McGrath, *Block Polymers*, Academic Press, New York (1977).

A. Opschoor, *Conformations of Polyethylene and Polypropylene*, Gordon and Breach, New York (1970).

F. Rodriguez, *Principles of Polymer Systems*, 2nd ed., McGraw-Hill, New York (1982).

I. D. Rubin, *Poly(One-Butene), Its Preparation and Properties*, Gordon and Breach, New York (1968).

A. G. Strots, *Polyolefins: Modifications of Structure and Properties*, Halsted, New York (1969).

13 | Polymeric Hydrocarbons with Pendant Groups

13.1 Polystyrene

Amorphous atactic polymers with bulky pendant groups, such as polystyrene (PS), which has the following repeating unit,

$$+ CH_2 - CH +$$

are less flexible at ordinary temperatures than low-density polyethylene (LDPE) or high-density polyethylene (HDPE). In addition to its steric hindrance effect, the flat phenyl group in PS also increases the intermolecular forces because of its weak polarity. However, the kinetic energy at higher temperatures exceeds the weak intermolecular forces, so this commercial polymer is a readily moldable thermoplastic.

PS has a high index of refraction (1.592) and hence has excellent transparency to visible light. PS is a brittle polymer with a glass transition temperature T_g of 100 °C, a heat deflection temperature of 90 °C, and a solubility parameter of 9.1 H.

As shown in Table 13.1, commercial PS has a specific gravity of 1.04 and a coefficient of linear expansion of 7.5×10^{-5} cm/cm · °C. It has a tensile strength of 41,000 kPa and an elongation of 1.5% at the yield point. It has a dielectric constant of 2.5 and excellent nonconductive electric properties.

Table 13.1   Polystyrene (PS)

Thermal and physical properties of plastics (typical values)	Unfilled PS	Impact PS	30%-Glass-filled PS
Heat deflection temperature @ 1820 kPa (°C)	90	90	105
Maximum resistance to continuous heat (°C)	75	70	95
Coefficient of linear expansion cm/ cm · °C $\times$ 10^{-5}	7.5	8.0	4.0
Compressive strength (kPa)	89,600	45,000	103,400
Flexural strength (kPa)	82,700	50,000	117,000
Impact strength (Izod: cm · N/ cm of notch)	21	80	80
Tensile strength (kPa)	41,000	41,000	82,000
Elongation (%)	1.5	3	1
Hardness: Rockwell	M60	M35	M60
Specific gravity	1.04	1.04	1.2

PS is soluble in aromatic hydrocarbon solvents and is resistant to aqueous solutions of nonoxidizing acids, alkalis, and salts. It is attacked by chlorine and oxidizing acids, such as 25% nitric and 95% sulfuric acids. In the absence of flame retardants, this aromatic hydrocarbon polymer burns readily and produces considerable amounts of black smoke. The latter is due to the formation of highly colored cyclic byproducts as the PS is burned.

The inherent brittleness of PS has been overcome by grafting with other monomers and blending with elastomers, including natural rubber (NR), and by copolymerization of styrene with flexibilizing monomers, such as butadiene. The specific gravity may be reduced by foaming (Styrofoam), and the specific gravity may be increased by the addition of fillers with high density. PS reinforced with 30% fibrous glass has a heat deflection temperature of 105 °C and a coefficient of linear expansion of 4.0 $\times$ 10^{-5} cm/cm · °C.

In contrast to the commercial at-PS, which is amorphous and transparent, it-PS is a very brittle crystalline translucent polymer which exists as a helix with three repeating units per turn. The crystalline polymer has a high specific gravity (1.12) and a high melting point T_m (230 °C).

13.2   Styrene Copolymers

The most widely used rigid copolymer of styrene is the random copolymer of styrene (70%)-acrylonitrile (30%) (SAN), which has randomly placed repeating units. This transparent copolymer is more ductile and has a higher

Table 13.2 ABS

Thermal and physical properties of plastics (typical values)	Extrusion grade	20% Glass-reinforced
Heat deflection temperature @ 1820 kPa (°C)	90	100
Maximum resistance to continuous heat (°C)	80	90
Coefficient of linear expansion cm/cm · °C $\times$ 10^{-5}	9.5	2.0
Compressive strength (kPa)	48,000	96,000
Flexural strength (kPa)	62,000	103,000
Impact strength (Izod: cm · N/ cm of notch)	320	53
Tensile strength (kPa)	34,000	75,000
Elongation (%)	60	5
Hardness: Rockwell	R60	M85
Specific gravity	1.03	1.2

heat deflection temperature (100 °C) than PS. There is a strong tendency for alternation in the copolymer. However, when large proportions of acrylonitrile are used, some of the acrylonitrile may be present as a block. These blocks cyclize when heated, and thus molded specimens may become yellow or dark.

The corrosion resistance of SAN is similar to that of PS, but the cyano pendant groups can be hydrolyzed by hot acids or alkalis. SAN has a solubility parameter of about 10 H and is more resistant than PS to aliphatic hydrocarbon liquids, such as those present in gasoline.

Most acrylonitrile-butadiene-styrene terpolymer (ABS) is produced as a graft of SAN onto a butadiene polymer backbone. This graft copolymer may be blended with more SAN or acrylonitrile elastomer (NBR) to improve its properties. ABS is more ductile than SAN. The T_g and the heat deflection temperature of ABS vary with the composition, and ABS may have one set of values for the PBD domains and another set for the SAN matrix. The permeabilities of ABS to oxygen, nitrogen, and carbon dioxide are much less than those of HDPE.

As shown in Table 13.2, the heat deflection temperature of ABS is increased by the incorporation of fibrous glass. The copolymer reinforced with 20% fibrous glass has a lower coefficient of linear expansion (2.0 $\times$ 10^{-5} cm/cm · °C) and higher compressive, flexural, and tensile strengths than unfilled SAN. These improvements are related to the amount of fibrous glass present in the composite.

Random copolymers of styrene and butadiene with high proportions of styrene are amorphous and transparent. These copolymers contain the following repeating units:

$$+CH_2-CH\!+\!CH_2-CH=CH-CH_2\!+$$

These commercial copolymers are more flexible than SAN and have lower T_g and lower heat deflection temperature values than commercial PS. These copolymers are resistant to aqueous solutions of most nonoxidizing acids, alkalis, and salts. Copolymers of styrene and butadiene are soluble in aliphatic hydrocarbon liquids but are insoluble in polar liquids, such as ethanol.

The copolymer of styrene and methyl methacrylate is a transparent, amorphous product which has glass-like optical properties.

13.3   Substituted Styrene Polymers

Polymers of alpha methylstyrene are usually low-molecular-weight and have the following repeating unit:

$$+CH_2-\overset{\overset{\displaystyle CH_3}{|}}{C}+$$

These polymers, which are commercially available, are not thermally stable and decompose to produce the monomer when heated. Copolymers of alpha methylstyrene with methyl methacrylate or styrene are transparent plastics with heat deflection temperatures greater than that of PS.

Because of increased steric hindrance, the polymers of *p*-chlorostyrene have higher heat deflection temperatures than PS. Polydichlorostyrene (Styramic) has a heat deflection temperature of 120 °C and a specific gravity of 1.4.

Linear alkyl groups in the para position reduce the intermolecular forces of polymers of styrene. Hence the heat deflection temperatures of poly-*p*-alkylstyrenes are lower than that of PS, and these heat deflection values decrease as the size of the alkyl group increases.

The specific gravity decreases as the size of the alkyl substituent increases, e.g., the specific gravity of poly-*p*-methylstyrene is 1.01, compared with 1.05 for PS.

Branched substituents on the nucleus of PS impede the rotation but do not decrease the T_g to any great extent. The solubility parameter decreases as the size of the substituent alkyl groups increases. Thus although PS is not soluble in aliphatic hydrocarbon liquids, poly-p-cyclohexylstyrene is soluble and serves as a viscosity index improver for lubricating oils.

13.4 Polyvinyl Chloride and Chlorinated Polyvinyl Chloride

Because of the presence of large polar pendant groups, which restrict molecular motion, unplasticized polyvinyl chloride (PVC), which has as the following repeating unit,

$$+ CH_2 - CH +$$
$$| $$
$$Cl$$

is a hard, rigid, brittle polymer with a low degree of crystallinity. In spite of the presence of some isotactic sequences and a small degree of crystallinity (5%), PVC is a transparent plastic. The presence of about 16 branches per molecule decreases the crystallinity of commercial PVC.

Commercial PVC is atactic and has a relatively high T_m (173 °C), which is 100 °C lower than that of it-PVC. Because of weak linkages and the presence of some unsaturation, commercial PVC tends to discolor and decompose at normal processing temperatures. More-stable polymers are obtained by the addition of stabilizers or by post-chlorination. Because of the presence of the chlorine pendant group, unplasticized PVC has a higher coefficient of friction than HDPE.

Chlorinated PVC (CPVC), which is available commercially, has a higher heat deflection temperature (105 °C) than rigid PVC (75 °C). CPVC has a higher specific gravity (1.55) than rigid PVC (1.4). Chlorination reduces the permeability of PVC to gases, such as carbon dioxide, oxygen, and nitrogen, as well as improving the stability of the polymer.

Rigid PVC has an index of refraction of 1.52 and low water absorption (0.5%). As shown in Table 13.3, the coefficients of linear expansion of rigid PVC and CPVC are 6.0 and 7.0 $\times$ 10^{-5} cm/cm $\cdot$ °C, respectively.

PVC and CPVC are resistant to nonoxidizing acids, alkalis, and salts. They are also more resistant than HDPE or PS to weak oxidizing acids, such as 10% nitric acid at room temperature. PVC has a solubility parameter of 9.5 H. Because of its moderate degree of crystallinity and relatively strong intermolecular forces, it is difficult to dissolve. It is soluble in cyclohexanone

Table 13.3   Polyvinyl chloride (PVC)

Thermal and physical properties of plastics (typical values)	Rigid PVC	Plasticized PVC	Chlorinated PVC (CPVC)
Heat deflection temperature @ 1820 kPa (°C)	75	—	105
Maximum resistance to continuous heat (°C)	60	35	90
Coefficient of linear expansion cm/cm · °C × 10^{-5}	6.0	12.5	7.0
Compressive strength (kPa)	68,000	6,000	103,000
Flexural strength (kPa)	89,000	—	103,000
Impact strength (Izod: cm · N/cm of notch)	27	—	—
Tensile strength (kPa)	44,000	10,000	55,000
Elongation (%)	50	200	5
Hardness: Rockwell	—	—	R120
Specific gravity	1.4	1.3	1.55

($\delta = 9.8$ H) and tetrahydrofuran ($\delta = 9.5$ H). PVC is not affected by strong polar solvents, such as ethanol; CPVC dissolves in acetone more readily than PVC.

Rigid PVC has been available commercially for over half a century, but the unstabilized polymer could not be molded or extruded until plasticizers were incorporated into it in the 1930s. The solubility parameters of commercial PVC plasticizers such as dioctylphthalate (DOP) are similar to those of PVC.

Plasticizers, which are essentially nonvolatile solvents, weaken the intermolecular forces and reduce crystallinity in PVC. A relatively stable suspension (called a *plastisol*) of finely divided PVC in a liquid plasticizer, can be poured into a mold and heated at about 175 °C to produce a solid flexible plastic as a result of fusion of the plasticizer in the PVC.

Smaller amounts (less than 10%) of plasticizers, actually, make PVC more rigid (antiplasticization), but much larger amounts (40%) produce a flexible plastic with a low T_g which can be readily molded and extruded.

As shown in Table 13.3, plasticized PVC has a higher coefficient of linear expansion (12.5 × 10^{-5} cm/cm · °C), lower resistance to temperature, lower strength, and higher elongation than rigid PVC. Plasticized PVC has good nonconductive electric properties and is slightly more permeable to gases than rigid PVC. Plasticized PVC is resistant to nonoxidizing acids, alkalis, and salts, as well as some polar solvents. However, some plasticizer may be extracted when plasticized PVC is in contact with polar solvents over a long period of time.

Because of the presence of the chlorine atom in every repeating unit, PVC does not burn as readily as HDPE. When used as a plasticizer, tricresyl phosphate also contributes to flame retardancy. In contrast, the organic ester plasticizers, such as DOP, contribute to the combustibility of plasticized PVC. Tricresyl phosphate is more toxic than organic esters, and DOP is more toxic than aliphatic esters, such as dioctyl adipate.

13.5 Copolymers of Vinyl Chloride and of Vinylidene Chloride

Copolymers of vinyl chloride are less crystalline and may be more flexible than unplasticized PVC. The random copolymer of vinyl chloride (87%) and vinyl acetate (13%) (Vinylite) contains the following repeating units:

$$\left(\!CH_2\!-\!\underset{\underset{Cl}{|}}{CH}\!\right)_{\!x}\!\!\left(\!CH_2\!-\!\underset{\underset{\underset{O}{\|}}{O-C-CH_3}}{CH}\!\right)_{\!y}$$

Vinylite is the most widely used vinyl chloride copolymer. This copolymer is not as strong nor as resistant to corrosives as rigid PVC but is more readily processed. PVC is flexibilized by blending it with elastomers (impact modifiers). These ductile blends of PVC and impact modifiers are more widely used when impact resistance is essential.

Because of the presence of polar pendant groups, the partially saponified vinyl acetate copolymer and copolymers of vinyl chloride, vinyl acetate, and maleic anhydride have better adhesion to metals than PVC.

Copolymers of vinyl chloride with small amounts of vinylidene chloride contain the following repeating units:

$$\left(\!CH_2\!-\!\underset{\underset{Cl}{|}}{CH}\!\right)_{\!x}\!\!\left(\!CH_2\!-\!CCl_2\!\right)_{\!y}$$

These copolymers are more flexible and more readily soluble than PVC.

Polyvinylidene chloride (PVDC) is a crystalline polymer with a low T_g and a low T_m. This homopolymer tends to degrade at elevated temperatures and is difficult to process.

The most widely used vinylidene chloride polymer is a copolymer of vinylidene chloride (90%) and vinyl chloride (10%) (Saran). This copolymer has a lower T_g and a lower T_m than the homopolymer. The copolymer, like the homopolymer, has a high specific gravity (1.8) and a very low permeability to gases and vapors. Films of this copolymer (Saran Wrap) are transparent.

Copolymers of vinylidene chloride and acrylonitrile are also transparent. These amorphous copolymers are also resistant to permeation by gases.

13.6 Fluorine-Containing Polymers and Copolymers

Polyvinyl fluoride (PVF) (Tedlar) has the following repeating unit:

$$-\!\!\!+\!CH_2-\underset{\underset{F}{|}}{CH}+\!\!\!-$$

PVF has a greater tendency to crystallize than PVC because F has a van der Waals radius near that of H. PVF has a melting point of 200 °C and a tendency to degrade at typical processing temperatures which are above 150 °C. PVF has a heat deflection temperature of 90 °C, a specific gravity of 1.4, and a coefficient of linear expansion of 10×10^{-5} cm/cm · °C. Because of the large number of fluorine atoms, which provide a screening effect for the carbon-carbon backbone, fluorinated polymers have excellent lubricity. PVF films have excellent resistance to weathering and are widely used as films and coatings.

Polyvinylidene fluoride (PVDF) (Kynar) has the following repeating unit:

$$-\!\!\!+\!CH_2-CF_2+\!\!\!-$$

PVDF is a polymer with a high degree of crystallinity, a T_g of -40 °C, a heat deflection temperature of 80 °C, and a melting point between 158 and 197 °C, depending on the crystalline form that is present. It may be processed without degradation to produce sheets or tubing. PVDF film has piezoelectric properties. PVDF is resistant to most acids, alkalis, salts, and solvents. It is soluble in dimethylacetamide (DMAc). It has a specific gravity of 1.76.

The random copolymer of vinylidene fluoride and hexafluoropropylene (Viton) has the following repeating units:

$$-\!\!\!+\!CF_2-\underset{\underset{F}{|}}{\overset{\overset{F_3C}{|}}{C}}\!+_x\!\!+\!CH_2-CF_2+_y$$

This copolymer is an amorphous elastomer. Because of its outstanding resistance to heat, solvents, and corrosives, and in spite of its brittleness at low temperatures, this copolymer is used for gaskets and O-rings.

Polychlorotrifluoroethylene (PCTFE) (Kel-F) has the following repeating unit:

$$\begin{array}{cc} Cl & F \\ | & | \\ \!\!\!-\!\!(C\!-\!C)\!\!- \\ | & | \\ F & F \end{array}$$

PCTFE is an atactic crystalline polymer with a T_g of 50 °C, a heat deflection temperature of 100 °C, and a T_m of 220 °C. PCTFE has a specific gravity of 2.1 and a coefficient of linear expansion of 14×10^{-5} cm/cm · °C.

Because of the presence of the larger chlorine atom on the backbone which reduces intermolecular forces and the degree of crystallinity, this polymer is more readily processed and has a lower T_m than polytetrafluoroethylene (PTFE). Thin films of PCTFE are transparent and are resistant to corrosives and solvents. The coefficient of friction of PCTFE is low but somewhat higher than that of PTFE.

PTFE (Teflon) has the following repeating unit:

$$-(CF_2\!-\!CF_2)-$$

PTFE is a crystalline polymer consisting of twisted zigzag spirals with at least 13 repeating units per turn. This nonpolar polymer has a solubility parameter of 6.2 H, a high T_m (327 °C), and a heat deflection temperature of 121 °C.

PTFE is a tough, flexible polymer which retains its ductility at extremely low temperatures (-269 °C). The coefficient of friction of PTFE is the lowest of any known solid material (see Table 13.4). Films of PTFE can be bonded by adhesives to other surfaces if the polymer surface is treated with sodium. It also bonds to diamonds and graphite whose surfaces have been fluorinated. Liquid sodium removes fluoride ions from the surface and leaves free radicals on the polymer surface. PTFE is resistant to almost all corrosives and solvents, but it can be dissolved in hot perfluorinated kerosene. PTFE is difficult to mold or extrude.

The properties of these fluoroplastics are shown in Table 13.4.

13.7 Polymers and Copolymers of Acrylonitrile

Polyacrylonitrile (PAN) has the following repeating unit:

$$-(CH_2\!-\!\underset{\displaystyle CN}{CH})-$$

Table 13.4   Fluoroplastics

Thermal and physical properties of fluoroplastics (typical values)	PTFE	PCTFE	PVDF	PVF	PE-CTFE	PE-TFE
Heat deflection temperature @ 1820 kPa (°C)	100	100	80	90	115	120
Maximum resistance to continuous heat (°C)	250	200	150	125	100	160
Coefficient of linear expansion cm/cm · °C $\times 10^{-5}$	10	14	8.5	10	8	7
Compressive strength (kPa)	27,000	38,000	—	—	41,000	48,000
Flexural strength (kPa)	—	60,000	—	—	48,000	38,000
Impact strength (Izod: cm · N/cm of notch)	160	133	—	—	—	—
Tensile strength (kPa)	24,000	34,000	55,000	—	48,000	48,000
Elongation (%)	200	100	200	—	200	250
Hardness: Rockwell	D52	R80	R110	—	R95	R50
Specific gravity	2.16	2.1	1.76	1.4	1.7	1.7

Commercial PAN is normally produced as an atactic polymer with strong hydrogen-bonded intermolecular forces. Because of repulsion between cyano pendant groups and intermolecular hydrogen bonds, the molecule assumes a crystallizable rodlike conformation. The hydrogen bonds between the rodlike chains create bundles of these chains. PAN may be spun into strong fibers. It has a T_g of 104 °C.

At temperatures above 160 °C, PAN forms cyclic imines which dehydrogenate in the presence of oxygen to produce dark-colored heat-resistant ladder polymers with conjugated C=C and C=N bonds.

PAN has a solubility parameter of about 13 H and is soluble only in polar solvents, such as dimethylacetamide (DMAc). Because of PAN's high polarity, its fibers are difficult to dye. This difficulty has been overcome by producing copolymers of acrylonitrile with small amounts (4%) of more-hydrophilic monomers, such as N-vinyl-2-pyrrolidone (left), methacrylic acid (center), or 2-vinylpyridine (right), which have the following structures:

Fibers with more than 85% acrylonitrile (Acrylan) are called *acrylic* fibers. Those with less acrylonitrile are called *modacrylic* fibers. The comonomers with acrylonitrile in modacrylic fibers are typically vinyl chloride or vinylidene chloride.

Commercial barrier resins, which are used as packaging films and blown bottles, are produced by blending copolymers of acrylonitrile, ethyl acrylate, and butadiene with selected copolymers of acrylonitrile. These barrier resins have a T_g of about 125 °C, a coefficient of linear expansion of 6.7 $\times$ 10^{-5} cm/cm · °C, a heat deflection temperature of 77 °C, and an index of refraction of 1.511. These resins are resistant to nonoxidizing alkalis and acids and are decomposed by mineral acids.

13.8 Polymers and Copolymers of Acrylamide and Methacrylamide

Polyacrylamide has the following repeating unit:

$$+CH_2-CH+$$
$$|$$
$$C=O$$
$$|$$
$$NH_2$$

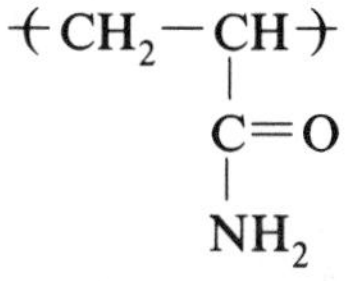

Because of the presence of the hydrophilic amide pendant group, polyacrylamide and polymethacrylamide are water-soluble polymers. These polymers have been used as flocculating agents and as agents for improving the flow of water in applications such as enhanced oil recovery. These polymers react with formaldehyde to produce polymers with methylol groups which are used in the textile industry. Copolymers of these monomers with acrylic or methacrylic acid are ionomers which improve the wet strength of paper by reacting with alum.

13.9 Polymers and Copolymers of Acrylic Acid and Methacrylic Acid

Polyacrylic acid has the following repeating unit:

$$+CH_2-CH+$$
$$|$$
$$C=O$$
$$|$$
$$OH$$

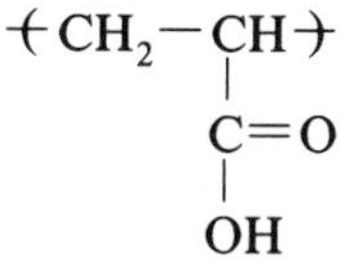

Polyacrylic acid and polymethacrylic acid and their sodium salts are water-soluble. Copolymers with small amounts of methacrylic acid and ethylene (ionomers), are moldable transparent resins.

13.10   Polymers and Copolymers of Alkyl Acrylates

Polymethyl acrylate (PMA) has the following repeating unit:

$$\text{+CH}_2\text{--CH+}$$
$$|$$
$$\text{C--O--CH}_3$$
$$||$$
$$\text{O}$$

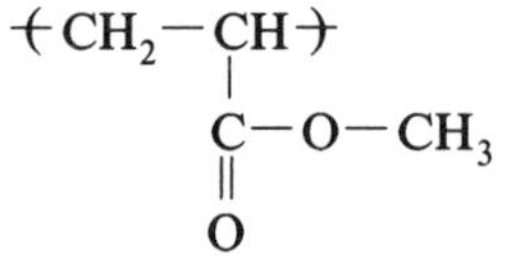

PMA is a tough, leathery resin with a low T_g and a solubility parameter of 10.5 H. In polymers of alkyl acrylates, the solubility parameter decreases as the size of the alkyl group increases. The flexibility also increases with the size of the pendant groups, but because of side chain crystallization, this tendency is reversed when the alkyl group has more than ten carbon atoms. Polyalkyl acrylates are readily hydrolyzed by alkalis to produce salts of polyacrylic acid. The copolymer of ethyl acrylate (95%) and chloroethyl vinyl ether (5%) is a commercial oil-resistant elastomer.

13.11   Polyalkyl-2-cyanoacrylates

Alkyl-2-cyanoacrylates are readily polymerized in the presence of weak bases, such as water. Because of the presence of the strong polar cyano pendant group, these polymers (e.g., Super Glue) are excellent adhesives. Polybutyl-2-cyanoacrylate is tolerated by the body better than its lower alkyl homologues. Hence polybutyl-2-cyanoacrylate is used as an adhesive aid to stop bleeding in some surgical operations.

13.12   Polymers and Copolymers of Alkyl Methacrylates

Polymers of alkyl methacrylates have the following repeating unit:

$$\text{CH}_3$$
$$|$$
$$\text{+CH}_2\text{--C+}$$
$$|$$
$$\text{C--OR}$$
$$||$$
$$\text{O}$$

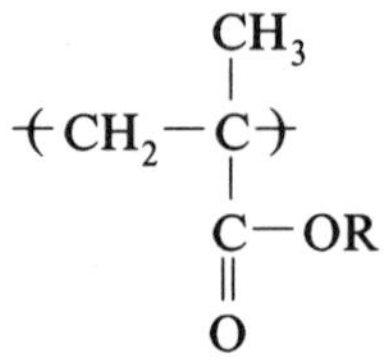

They are atactic amorphous polymers which have good light transparency (92%) and yield transparent moldings and films. As was noted for polyalkyl acrylates, the solubility parameters decrease as the size of the alkyl groups increases. The flexibility also increases as one goes from polymethyl methacrylate (PMMA) to polyaryl methacrylate and then decreases as the size of the alkyl group is further increased.

The presence of two substituents on every alternate carbon atom restricts chain mobility, so polyalkyl methacrylates are less flexible than the corresponding polyalkyl acrylates. Also, the presence of an alpha alkyl group increases (compared with the corresponding polyalkyl acrylates) the stability of polyalkyl methacrylates to light and chemical degradation.

Unlike the polyalkyl acrylates, which are thermally degraded by random chain scission, polyalkyl methacrylates unzip when heated, and excellent yields of the monomers are produced when the polymers of the lower homologues are heated. When higher homologues are heated, there is also some thermal degradation of the alkyl substituents.

The most widely used acrylic plastics are PMMA (Lucite) or copolymers of methyl methacrylate with small amounts (2 to 18%) of methyl or ethyl acrylate (Plexiglas). These commercial products, which are available as sheets and as molding powders, have a specific gravity of about 1.2, a heat deflection temperature of about 95 °C, a refractive index of about 1.5, and a water absorption of 0.2%. PMMA is more resistant to impact than PS or glass, but its scratch resistance is inferior to that of glass.

As shown in Table 13.5, PMMA has a coefficient of linear expansion of 7.0×10^{-5} cm/cm · °C. PMMA is resistant to nonoxidizing acids, alkalis, and salts at room temperature, but it is attacked by oxidizing acids at ordinary temperatures and by nonoxidizing acids and alkalis at elevated temperatures (90 °C). It is resistant to highly polar solvents, such as ethanol, but is soluble in less-polar solvents, such as toluene.

Table 13.5 Properties of Polymethyl Methacrylate (PMMA)

Thermal and physical properties of plastics (typical values)	PMMA
Heat deflection temperature @ 1820 kPa (°C)	95
Maximum resistance to continuous heat (°C)	75
Coefficient of linear expansion cm/cm · °C $\times$ 10^{-5}	7.0
Compressive strength (kPa)	103,000
Flexural strength (kPa)	96,000
Impact strength (Izod: cm · N/cm of notch)	21.4
Tensile strength (kPa)	65,000
Elongation (%)	4
Hardness: Rockwell	M80
Specific gravity	1.18

Cross-linked copolymers of hydroxyethyl methacrylate and methyl methacrylate are slightly swollen by water and are used as soft contact lenses.

13.13   Polyvinyl Acetate

Because of the presence of atactic acetyl groups, polyvinyl acetate (PVAc), whose repeating unit is as follows,

$$\text{--}CH_2\text{--}CH\text{--}$$
$$O\text{--}C\text{--}CH_3$$
$$\|$$
$$O$$

is a flexible, soft amorphous polymer with a T_g of 28 °C. The many branches present in PVAc contribute to some stiffness at room temperature. Because of the presence of the polar acetyl groups, PVAc is a good adhesive.

PVAc has a specific gravity of 1.2 and an index of refraction of 1.47. It has a solubility parameter of 9.5 H and is soluble in liquids with similar solubility parameter values, such as benzene, chloroform, and acetone.

13.14   Polyvinyl Alcohol and Polyvinyl Acetals

Polyvinyl alcohol (PVA) produced by the hydrolysis of PVAc has the following repeating unit:

$$\text{--}CH_2\text{--}CH\text{--}$$
$$OH$$

PVA is an atactic crystalline polymer. The hydroxyl groups, which are much smaller than the acetyl groups in PVAc, permit good packing of the polymer chains.

Because of strong hydrogen bonding, the completely hydrolyzed PVAc (pure PVA) is insoluble in water. Relatively pure PVA decomposes at temperatures below the high temperatures required for processing.

It is customary to stop the hydrolysis of PVAc before all the acetyl groups are removed. Thus the commercial product, with a degree of hydrolysis of about 88%, is readily soluble in water but is resistant to less polar solvents, such as benzene and gasoline. PVA fibers (Kuralon) are strong and insoluble in water because of a surface treatment with formaldehyde which reacts with the surface hydroxyl groups to produce polyvinyl formal on the polymer surface.

Polyvinyl acetals have the following repeating unit:

$$+CH_2-CH-CH_2-CH+$$

$$\underset{R}{\overset{\displaystyle O\diagdown \underset{\displaystyle C}{\overset{\displaystyle H}{|}} \diagup O}{}}$$

These products are produced by the reaction of partially hydrolyzed PVAc with aldehydes. The acetal rings on these random amorphous polymer chains restrict flexibility and increase the heat deflection temperature to a value higher than that of PVAc. The heat deflection temperature of polyvinyl formal is about 90 °C and is dependent on the specific composition of this complex polymer. Because of the presence of residual hydroxyl groups, commercial polyvinyl formal has a water absorption of about 1%. Polyvinyl formal has a T_g of 105 °C. It has a solubility parameter of about 10 H and is soluble in solvents with similar solubility parameters, such as acetone.

The most widely used polyvinyl acetal is polyvinyl butyral (PVB). This transparent amorphous plastic is used as a plasticized polymer in the inner lining of safety windshield glass (Saflex). Because of the presence of hydroxyl groups, the commercial product, which is produced from 75% hydrolyzed PVAc, has a T_g of about 49 °C and has excellent adhesion to glass.

13.15 Polyvinyl Methyl Ether

Polyvinyl methyl ether has the following repeating unit:

$$+CH_2-CH+$$
$$\overset{|}{OCH_3}$$

The commercial polymer is a soft, water-soluble, typically atactic flexible material with a T_g of -20 °C. This polymer is used for pressure-sensitive adhesive formulations. The polar methoxy groups keep the polymer chains apart and promote flexibility. Polymers with higher-molecular-weight alkoxy groups are also available commercially.

13.16 Polyvinyl-2-pyrrolidone

Polyvinyl-2-pyrrolidone (PVP) has the following repeating unit:

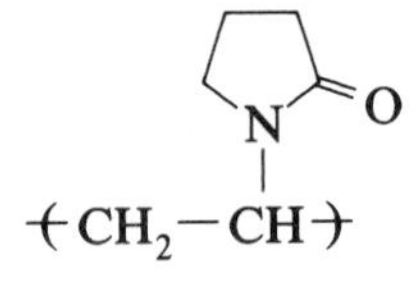

$$+CH_2-CH+$$

PVP is a flexible polymer which is soluble in water and other polar solvents. PVP has a T_g of 17.5 °C. Solutions of PVP are used in hair spray formulations and as blood extenders.

13.17 Poly-*N*-vinylcarbazole

Poly-*N*-vinylcarbazole (Luvican) has the following repeating unit:

$$+CH_2-CH+$$

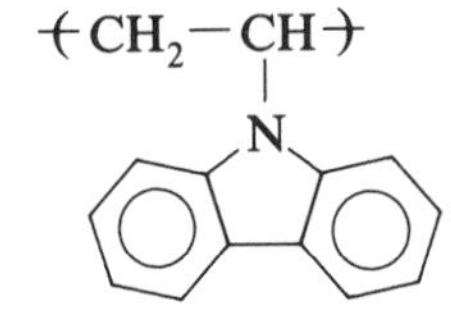

Poly-*N*-vinylcarbazole is a stiff polymer with a high T_g (150 °C). Because this polymer has excellent electric properties, it has been used as a capacitor dielectric.

13.18 Polychloroprene

Polychloroprene (Neoprene) has the following repeating unit:

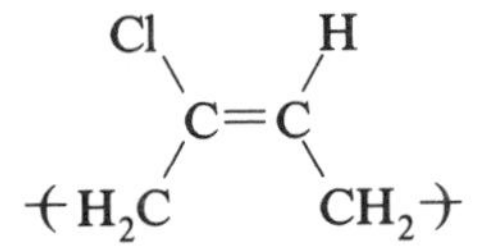

Neoprene is an oil-resistant elastomer with a T_g of −45 °C. It has an index of refraction of 1.558 and a coefficient of linear expansion of 60×10^{-5} cm/cm · °C.

13.19 References

R. B. Bishop, *Practical Polymerization for Polystyrene,* CBI, Boston, Mass. (1971).

R. H. Burgess, *Manufacture and Processing of PVC,* Macmillan, New York (1981).

L. Nass, *Encyclopedia of PVC,* Dekker, New York, Vol. 1 (1976), Vols. 2 and 3 (1977).

J. G. Pritchard, *Polyvinyl Alcohol,* Gordon and Breach, New York (1970).

H. Raech, *Allylic Resins and Monomers,* Krieger, Melbourne, Fla. (1965).

I. D. Rubin, *Poly(One-Butene), Its Preparation and Properties,* Gordon and Breach, New York (1968).

C. E. Schildknecht, *Allyl Compounds and Their Polymers: Including Polyolefins,* Wiley, New York (1973).

R. B. Seymour, *Plastics vs. Corrosives,* Wiley, New York (1982).

R. B. Seymour, *Modern Plastics Technology,* Reston Publishing, Reston, Va. (1975).

R. A. Wessling, ed., *Polyvinylidene Chloride,* Gordon and Breach, New York (1977).

14 | Aliphatic Polymers with Heteroatom Chains

14.1 Introduction

The polymer molecules described previously were catenated carbon chains with various pendant groups. The polymer molecules described in this chapter have other atoms in the polymer chain, either in addition to or in place of the carbon atoms.

14.2 Polyethers

Polyoxymethylene (POM), which is called polyacetal, is a crystalline polymer of formaldehyde which has the following repeating unit:

$$\text{-(}O - CH_2\text{)-}$$

Since this polymer unzips to produce formaldehyde at temperatures above 127 °C, the terminal hydroxyl end groups in the commercial polymer (Delrin) are acetylated (capped) by reaction with acetic anhydride. Because of the polar, short C—O bonds, which permit close packing, the degree of crystallinity in commercial POM is at least 70%. As shown in Table 14.1, POM has a specific gravity of 1.412 and a relatively high heat deflection temperature (125 °C). Because of good symmetry, the homopolymer POM has a relatively high melting point T_m of about 180 °C. The methylene and oxygen linkages in the chain promote flexibility and superior resistance to abrasion.

POM is available with molecular weights between 3 and 5×10^4. It is stable in air at temperatures up to 100 °C and exhibits good mechanical strength over a wide temperature range. POM has a solubility parameter of

Table 14.1 Acetals

Thermal and physical properties of plastics (typical values)	Homopolymer	Copolymer	25%-Reinforced copolymer
Heat deflection temperature @ 1820 kPa (°C)	125	110	160
Maximum resistance to continuous heat (°C)	100	100	125
Coefficient of linear expansion cm/cm · °C $\times$ 10^{-5}	10.0	8.5	5.0
Compressive strength (kPa)	106,110	110,320	117,215
Flexural strength (kPa)	96,530	89,635	193,060
Impact strength (Izod: cm · N/ cm of notch)	80.1	69.4	96.1
Tensile strength (kPa)	68,950	62,055	128,600
Elongation (%)	30	50	3
Hardness: Rockwell	M94	M78	M79
Specific gravity	1.412	1.41	1.61

11.1 H but, because of its high crystallinity, is not affected by many solvents at room temperature. It is resistant to polar solvents, such as ethanol, but is attacked by nonpolar solvents, such as benzene at 25 °C.

POM has a coefficient of linear expansion of 10.0×10^{-5} cm/cm · °C and high tensile, flexural, and compressive strengths. POM is attacked by acids but is resistant to alkalis and salts at room temperature.

Random acetal copolymers (Celcon) have the following repeating unit:

$$\{CH_2-O\}_x\{CH_2-CH_2-O\}_y$$

These copolymers are thermally stable, but they have slightly lower tensile strengths than the homopolymer. The copolymer has a lower heat deflection temperature (110 °C) and a low coefficient of friction and specific gravity. When reinforced by 25% fiberglass, the copolymer has a heat deflection temperature of 160 °C, a coefficient of linear expansion of 5.0×10^{-5} cm/cm · °C, a specific gravity of 1.61, and much higher tensile and flexural strengths than the unfilled copolymer.

Because of the additional methylene groups, polyethylene oxide (PEO, Polyox), which has the following repeating unit,

$$\{CH_2-CH_2-O\}$$

is more flexible than POM. This crystalline polymer is soluble in water and moderately polar organic solvents.

Polypropylene oxide is less hydrophilic than PEO. Block copolymers of propylene oxide and ethylene oxide (Pluronics) contain lyophilic and hydrophilic domains and are used as surface-active agents.

Polymers of epichlorohydrin and copolymers of epichlorohydrin with ethylene oxide are atactic, flexible, amorphous elastomers with the following repeating unit:

$$+ CH_2 - CH_2 - O +$$
$$\qquad\quad | $$
$$\qquad CH_2 - Cl$$

These ozone- and oil-resistant elastomers may be cross-linked by amines.

Poly-3,3-bischloromethyl oxetane (Penton) is a crystalline, flame-, thermal-, and solvent-resistant polymer with the following repeating unit:

$$\qquad\qquad CH_2Cl$$
$$\qquad\qquad\quad |$$
$$+ CH_2 - C - CH_2 - O +$$
$$\qquad\qquad\quad |$$
$$\qquad\qquad CH_2Cl$$

This high-melting (130 °C) polymer is resistant to oxidizing acids at temperatures up to 65 °C.

14.3   Polyolefin Polysulfides

Polyolefin polysulfides (Thiokol) are thio analogues of polyethers which are flexible, amorphous, oil-resistant elastomers. The number of sulfur links in the repeating unit, which is called *rank*, is always greater than two. The solvent resistance, resistance to gaseous permeation, and flexibility of these polyolefin sulfides increase with rank. The actual strength of polymers and copolymers of olefin sulfides varies. Many of these products contain the following repeating unit:

$$+ CH_2 - CH_2 - O - CH_2 - CH_2 - S_x +$$

Liquid polymers with terminal mercapto groups (LP-2) are obtained when solid polyolefin polysulfides are reduced. These liquid polymers may be oxidized to solid polymers in the presence of oxidizing agents, such as lead oxides. The liquid polymers are used in caulking compositions and as reactive flexibilizing agents in epoxy resins.

14.4   Aliphatic Polyesters

Aliphatic polyesters, such as polyethylene adipate, have the following repeating unit:

$$\text{+}O-CH_2CH_2O-COCH_2CH_2CH_2CH_2CO\text{+}$$

These polyesters were investigated by W. H. Carothers in the 1930's, but they were not commercialized at that time because of their high flexibility and low melting point (54 °C), which made them unsuitable for use as fibers that could withstand the heat of a flat iron. They are used as flexibilizers for other polymers and as reactants for polyurethanes. (PUs).

Polymers of diethylene glycol bis(allyl carbonate) (CR-39) are optically clear, infusible polymers which are used for lenses.

Polyglycolides have the following repeating unit:

$$\text{+}\underset{\underset{O}{\|}}{C}-CH_2-O\text{+}$$

They are flexible polymers which are used as sutures that are absorbed in the body.

14.5 Nylons

Hexamethylene adipamide (nylon 66), in which the first integer is equal to the number of carbon atoms in the diamine and the second integer is equal to the number of carbon atoms in the dicarboxylic acid, has the following repeating unit:

$$\text{+}\underset{\underset{O}{\|}}{C}\text{+}CH_2\text{+}_4-\underset{\underset{O}{\|}}{C}-\overset{\overset{H}{|}}{N}\text{+}CH_2)_6-\overset{\overset{H}{|}}{N}\text{+}$$

Nylon 66 is a polar, crystalline polyamide with a T_m of 265 °C. As shown in Table 14.2, nylon 66 has a heat deflection temperature of 75 °C and a coefficient of linear expansion of 8.0×10^{-5} cm/cm · °C.

Nylon 66 has a low coefficient of friction. In the presence of water, this value is lowered below that of graphite. The coefficient of friction in the presence of liquid paraffin is extremely low (0.13). Nylon 66 has excellent resistance to abrasion.

Nylon 66 has fair-to-good nonconductive electric properties, but these properties are diminished in the presence of moisture. Nylon 66 is the most widely used engineering plastic, but its principal use is as a fiber.

Like most other strong fibers, nylon 66 has a symmetrical structure which permits good fitting of the adjacent chains, and it also has hydrogen

Table 14.2 Nylons

Thermal and physical properties of plastics (typical values)	Nylon 66	Glass-filled Nylon 66	Nylon 6	Glass-filled Nylon 6
Heat deflection temperature @ 1820 kPa (°C)	75	250	80	210
Maximum resistance to continuous heat (°C)	120	140	125	130
Crystalline melting point (°C)	265	—	225	—
Coefficient of linear expansion (cm/cm · °C × 10^{-5})	8.0	2.0	8.0	3.0
Compressive strength (kPa)	103,500	207,000	96,500	131,000
Flexural strength (kPa)	103,500	276,000	96,500	207,000
Impact strength (Izod: cm · N/cm of notch)	80	106.7	160	160
Tensile strength (kPa)	82,750	172,000	62,055	172,000
Ultimate elongation (%)	30	3	—	—
Hardness: Rockwell	R120	M100	M119	M101
Specific gravity	1.2	1.4	1.15	1.4

bonds which assure strong intermolecular attractions among these chains. Nevertheless, because of chain flexibility related to the many methylene groups between the amide stiffening groups, nylon 66 has a low melt viscosity and may be readily extruded to produce unoriented filaments. These filaments are oriented by stretching (drawing) to produce strong fibers in which the chains are aligned so that the intermolecular forces are more effective.

The amide groups, which are water-sensitive, are also responsible for the good dyeability of nylon 66. Nylon 66 has a specific gravity of 1.2 and excellent strength properties. When nylon 66 is reinforced by fiberglass, the heat deflection temperature is increased to 250 °C, the coefficient of linear expansion is decreased to 2.0×10^{-5} cm/cm · °C, and the tensile strength is increased. In contrast, the impact strength of nylon 66 is improved by blending it with terpolymers of ethylene and propylene containing carboxylic acid groups.

Nylon 66 is resistant to aqueous salts and alkalies but deteriorates in the presence of mineral acids. It is resistant to nonpolar solvents, such as gasoline, but is softened by polar solvents, such as ethanol. Nylon 66 is soluble in formic and acetic acids and in phenol and cresols. In the absence of stabilizers, nylon 66 deteriorates when exposed to sunlight.

The glass transition temperature T_g of aliphatic nylons is low (40 to 70 °C) and is not affected to any great extent by composition. However, the T_m is inversely related to the number of methylene groups present in the repeating unit. Thus the T_m of nylon 66 is lower than that of nylon 46 (275 °C). The flexibility is increased and the water absorption is decreased as the number of methylene groups in the repeating unit increases. Because of the ability of the adjacent polymer chains to fit together better, the T_m of dyadic nylons with even numbers of carbon atoms is greater than that of comparable nylons with odd numbers of carbon atoms. Thus nylon 66 has a higher T_m than either nylon 56 or nylon 76.

Several commercially available monadic nylons have the following repeating unit:

$$\begin{matrix} H && O \\ | && \| \\ \end{matrix}$$
$$+\!N\!-\!CH_2\!\!\xrightarrow{}_x\!C\!+$$

The value of x is 1 in nylon 2, which is the lowest member of this series of monadic polyamides.

The T_m and the water absorption of nylon 3 are much lower than those of nylon 2, and the water absorption of nylon 4 is less than that of nylon 3.

Because the repeating units of polyamides with an even number of methylene groups fit together better than those of polyamides with an odd number of methylene groups, the T_m of nylons with even numbers of carbon atoms is greater than that of comparable nylons with odd numbers of carbon atoms. The T_m of nylon 4 is less than that of nylon 3.

Nylon 6 (polycaprolactam, Perlon) is the most widely used monadic nylon. It has a lower melting point (225 °C) than nylon 5 (270 °C), which is not available commercially. As shown in Table 14.2, the coefficient of linear expansion of nylon 6 is the same as that of nylon 66. However, its heat deflection temperature (80 °C) is slightly higher than that of nylon 66 (75 °C). Nylon 7 (polyenantholactam) has a slightly higher T_m (227 °C), a slightly greater flexibility, and a lower water absorption than nylon 6.

Nylon 12 has a lower T_m (185 °C), greater flexibility, and lower water absorption than nylon 6. Cycloaliphatic polyimides, such as those produced by the condensation of 1,4-bis(aminomethyl) cyclohexane and suberic acid, have relatively low water absorption and excellent nonconductive electric properties.

The aromatic polyamides (aramids) are high-melting, stiff, relatively insoluble polymers. The polymer obtained by the condensation of hexamethylenediamine and terephthalic acid has a melting point of 370 °C and is soluble only in concentrated H_2SO_4 and hot dimethylacetamide (DMAc). As shown in Table 14.3, this linear aramid (poly-*p*-phenylene phthalamide, Kevlar) has a heat deflection temperature of 260 °C and outstanding mechanical properties. Poly-*m*-phenylene phthalamide (Nomex) exists as a zigzag chain.

The T_m's of aliphatic and aromatic (aramids) nylons are tabulated in Table 14.4.

Aromatic polyimides have the following repeating unit:

$$\left. \left(N \underset{\underset{O}{\overset{\overset{O}{\parallel}}{\overset{C}{\underset{C}{\parallel}}}}{\overset{\overset{O}{\parallel}}{\overset{C}{\underset{C}{\parallel}}}} O \underset{\overset{\overset{O}{\parallel}}{\overset{C}{\underset{C}{\parallel}}}}{\overset{\overset{O}{\parallel}}{\overset{C}{\underset{C}{\parallel}}}} N - R \right) \right.$$

where R is the aliphatic or aromatic group. These polymers are often processed as polyamic acids, which are soluble in N-methylpyrrolidone and DMAc. The polyimides, obtained by heating the polyamic acids, are high-melting, intractable polymers which retain their good mechanical properties at temperatures above 350 °C in air. Some properties of a typical polyimide are shown in Table 14.5.

Polyamide imides and polyester imides are more flexible and more readily processed than polyimides.

14.6 Polyurethanes and Polyureas

PUs have the following repeating unit:

$$\left(O - R - O - \underset{\overset{O}{\parallel}}{C} - \underset{\overset{H}{\mid}}{N} - R - \underset{\overset{H}{\mid}}{N} - \underset{\overset{O}{\parallel}}{C} \right)$$

Table 14.3 A Typical Aramid

Thermal and physical properties of plastics (typical values)	Aramid
Heat deflection temperature @ 1820 kPa (°C)	260
Maximum resistance to continuous heat (°C)	150
Coefficient of linear expansion cm/cm · °C $\times$ 10^{-5}	2.6
Compressive strength (kPa)	206,850
Flexural strength (kPa)	172,325
Impact strength (Izod: cm · N/cm of notch)	74.5
Tensile strength (kPa)	120,000
Elongation (%)	5
Hardness: Rockwell	E90
Specific gravity	1.2

Table 14.4 Approximate Crystalline
Melting Points of Selected Polyamides
(Nylons)

Aliphatic nylons	
Nylon	Melting point (°C)
3	320
4	265
5	270
6	225
7	227
8	195
9	200
10	185
11	190
12	185
46	275
56	225
66	265
410	240
510	190
610	230

Aromatic nylons (terephthalamides)	
Diamine	Melting point (°C)
1,2-Ethylene	460
1,3-Trimethylene	400
1,4-Tetramethylene	440
1,5-Pentamethylene	350
1,6-Hexamethylene	370

Table 14.5 Polyimides

Thermal and physical properties of plastics (typical values)	Thermoplastic	50% glass-filled thermoset
Heat deflection temperature @ 1820 kPa (°C)	315	350
Maximum resistance to continuous heat (°C)	300	325
Coefficient of linear expansion cm/cm · °C × 10^{-5}	5.0	1.3
Compressive strength (kPa)	241,300	234,400
Flexural strength (kPa)	172,400	144,800
Impact strength (Izod: cm · N/cm of notch)	80	294
Tensile strength (kPa)	96,500	44,000
Elongation (%)	8	0.5
Hardness: Rockwell	E60	M118
Specific gravity	1.4	1.6

The properties of these versatile polymers may be modified by changes in composition. As is the case for polyamides, PUs with an even number of carbon atoms between characteristic groups in the chain have a higher T_m than the corresponding polymers with an odd number of carbon atoms between such groups. Thus the polymer produced by the reaction of a diisocyanate with four carbon atoms and a glycol with four carbon atoms has a higher T_m (190 °C) than one produced from a diisocyanate with five carbon atoms and a glycol with four carbon atoms (159 °C). The T_m and the water absorption decrease and the flexibility increases as the number of carbon atoms between characteristic groups is increased.

The T_m's of PUs are lower than those of corresponding nylons, but they have greater affinity for dyes and have lower water absorption. PUs are usually soluble in formic acid and phenol. The T_m values of selected PUs are shown in Table 14.6.

PUs are resistant to aqueous alkalies, salts, and acids. Because of the availability of hydrogen bonds, PUs bond well to other "polar-like" surfaces, including glass.

As in the case of nylons, the flexibility of PUs is increased as the number of methylene groups is increased, and the rigidity is related to the number of stiffening groups, such as phenylene groups in the chain. As the number of methylene units increases (Table 14.6), the T_m decreases. The T_m generally increases as the flexible units are replaced by nonflexible units, such as phenylenes and piperazines. Thermal and physical properties of aliphatic PUs are shown in Table 14.7.

Table 14.6 Crystalline Melting Points for Selected Polyurethanes

$$\left\{ O \left\{ CH_2 \right\}_m O - \overset{\overset{\displaystyle O}{\|}}{C} - \overset{\overset{\displaystyle (H)}{|}}{N} - R - \overset{\overset{\displaystyle (H)}{|}}{N} - \overset{\overset{\displaystyle O}{\|}}{C} \right\}$$

m	R (or diamine fraction)	Melting Point (°C)
2	1,2-Ethylene	225
2	1,4-Hexamethylene	180
6	1,6-Hexamethylene	170
2	N⌯N (piperazine)	250
6	N⌯N (piperazine)	120

Table 14.7   Aliphatic Polyurethanes (PUs)

Thermal and physical properties of plastics (typical values)	Cast PU	Elastomer PU
Maximum resistance to continuous heat (°C)	70	65
Coefficient of linear expansion cm/cm · °C $\times$ 10^{-5}	15.0	15.0
Compressive strength (kPa)	138,000	138,000
Flexural strength (kPa)	20,680	27,500
Impact strength (Izod: cm · N/cm of notch)	1335.0	No break
Tensile strength (kPa)	6,890	7,250
Elongation (%)	200	150
Specific gravity	1.2	1.15

Polyureas have the following repeating unit:

$$\left. \begin{array}{ccc} & \overset{\displaystyle H}{|} & & \overset{\displaystyle H}{|} \\ -N & - & C & - N - R \\ & & \overset{\displaystyle \|}{O} & \end{array} \right.$$

These high-melting products are produced by the reaction of diamines and diisocyanates.

14.7   Polypeptides—Proteins

Proteins, which have the following repeating unit,

$$\left. \begin{array}{ccc} \overset{\displaystyle H}{|} & \overset{\displaystyle R}{|} & \overset{\displaystyle O}{\|} \\ N & - CH - & C \end{array} \right.$$

are copolymers of some 20 different alpha amino acids. Since all the amino acids except glycine (H_2N-CH_2-COOH) are substituted 1-alpha-amino-car-boxylic acids, these polypeptides are also optically active. Many biochemists use the term *polypeptide* to describe low- and moderate-molecular-weight polyamides and reserve the term *protein* for the high-molecular-weight products.

Biochemists use the terms *primary* and *secondary structure* to designate configurations and conformations of proteins, respectively.

Proteins occur as helices (alpha structures) and pleated sheets (beta structures). The alpha structures are dependent on intramolecular hydrogen bonds, whereas the beta structures are dependent on intermolecular hydrogen bonding.

In nature, extended helical conformations appear to be utilized in two major ways: to provide linear systems for the storage, duplication, and transmission of information (DNA, RNA), and to provide inelastic fibers for the generation and transmission of forces (F-actin, myosin, and collagen).

Examples of the various helical forms found in nature are the single helix (RNA), the double helix (DNA), the triple helix (collagen fibrils), and complex multiple helices (myosin, F-actin). Generally, these single and double helices are fairly readily soluble in dilute aqueous salt solution. The triple and complex helices are soluble only if the secondary bonds are broken.

There are a number of examples in which polypeptide chains are arranged in parallel rows joined by covalent crosslinks at regular intervals, leading to network structures. This is called the *alpha arrangement*; these polymers tend to be similar to synthetic crosslinked polymers in that they cannot be solubilized without covalent bond breakage, and they tend to be good structural materials. The alpha keratin of wool consists of parallel polypeptide alpha helices linked by disulphide bonds. If alpha keratin is subjected to tension in the direction of the helix axes, the hydrogen bonds parallel to the axes are broken, and the structure can be irreversibly elongated to an extent of about 100%. The relation between the mechanical property and the chemical structure is diagrammed in Figure 14.1.

The *beta arrangement*, or pleated sheet conformation (see Figure 14.2), is predominant when small pendant groups are present in the chain, as in silk fibroin. The silk fibroins, which are spun by various species of silkworms, are monofilament polypeptides with extensive secondary interchain bonding. The crystalline portion of the fibroin is a polymer of a hexapeptide. The

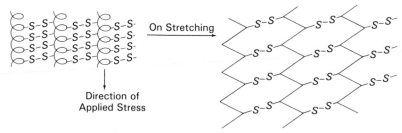

Figure 14.1   Stretching of alpha keratin.

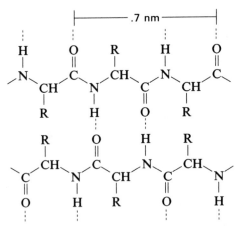

Figure 14.2 Beta arrangement or pleated sheet conformation of proteins.

polypeptides are arranged in antiparallel beta pleated sheets which allow multiple hydrogen bonding at right angles to the polypeptide chains.

This structure, because of the antiparallel packing, gives rise to alternate 3.5 and 5.7 Å spacings between layers, repeating many times over, in the plane of the sheet. The arrangement is shown in Figure 14.3.

The polypeptide chains are virtually fully extended. There is a little puckering which allows for optimum hydrogen bonding, and hence the structure is inextensible in the direction of the polypeptide chains. On the other hand, there are less specific, hydrophobic (London, dispersion) forces between the sheets, and this permits flexibility.

Thus the crystalline segments of silk fibroin exhibit three types of bonding: covalent bonding, hydrogen bonding, and hydrophobic bonding. The physical properties of the crystalline regions are in accord with these bonding types.

The crystalline regions in the polymers are interspersed with amorphous

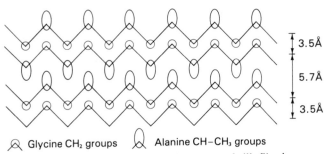

Figure 14.3 Three-dimensional structure of silk fibroin.

regions in which glycine and alanine are replaced by other amino acids with bulkier side chains. This prevents ordered arrangements in these regions.

The term *tertiary structure* is employed to designate the shape or folding of polymers. The integrity of globular polymers is dependent on the sum of the effects of secondary interactions, i.e., hydrogen bonds, hydrophobic bonds, and salt linkages. Consequently, solvents which break such bonds lead to a randomization of the polypeptide chains which is known as *denaturation*. Denaturation may be accompanied by the globular polymer's adopting a random coil conformation in the denaturing solvent, by the formation of a gel, or by precipitation, depending on the composition of the denaturing solvent.

14.8   Polysaccharides

The most important polysaccharides are cellulose and starch. These may be hydrolyzed by acids or enzymes to lower-molecular-weight carbohydrates (oligosaccharides) and finally to D-glucose. The latter is the building block for most carbohydrate polymers. Cellobiose and maltose, which are the repeating units in cellulose and starch, respectively, are disaccharides, consisting of two molecules of D-glucose joined together through carbon atoms 1 and 4. The D-glucose units in cellobiose are joined by beta acetal linkages, whereas those in maltose are joined by alpha acetal linkages, as shown in Figure 14.4.

Cellulose is the most abundant naturally occurring organic material. It acts as the main structural component in higher plants. The purest form of cellulose is derived from the seed hairs of the cotton plant, which contains 95% cellulose.

Because of the beta configuration, the glucose units in cellulose effectively alternate throughout the chain. This permits the individual cellulose chains to align themselves side by side, taking advantage of intrachain and interchain hydrogen bonding. The intermolecular bonding is so strong that cellulose is insoluble in water, whereas the linear amylose chains (alpha-1 → 4 linked) are water-soluble. Also, although amylose is flexible, cellulose is rigid.

Highly ordered crystalline cellulose has a density as high as 1.63 g·cm^{-3}, whereas highly disordered amorphous cellulose has a density as low as 1.47 g·cm^{-3}. High-molecular-weight native cellulose, which is insoluble in 17.5% aqueous sodium hydroxide solution, is called alpha cellulose. The fraction that is soluble in 17.5% sodium hydroxide solution but insoluble in 8% solution, is called beta cellulose, and that which is soluble in 8% sodium hydroxide solution is called gamma cellulose.

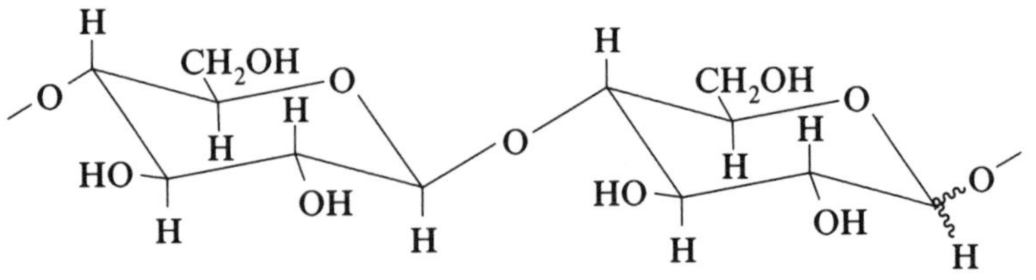

Cellobiose repeating unit

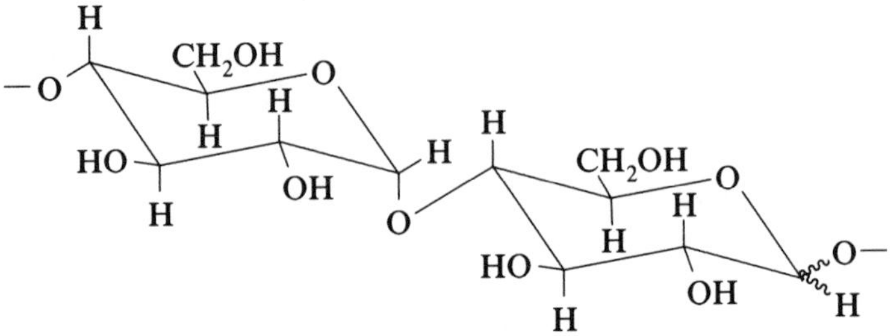

Maltose repeating unit

Figure 14.4 Repeating units of cellobiose and maltose.

Strong caustic solutions penetrate the crystal lattice of alpha cellulose and produce an alkoxide called alkali, or soda, cellulose. Mercerized cotton is produced by aqueous extraction of the sodium hydroxide from alkali cellulose fibers. Cellulose ethers and cellulose xanthate are produced by reactions of alkyl halides or carbon disulfide, respectively, with the alkali cellulose.

Most linear cellulosics may be dissolved in solvents capable of breaking the strong hydrogen bonds. These solvents include aqueous solutions of inorganic acids, zinc chloride, lithium chloride, dimethyl dibenzyl ammonium hydroxide, and cadmium or copper ammonia hydroxide (Schweizer's reagent). Cellulose is also soluble in hydrazine, dimethyl sulfoxide in the presence of formaldehyde, and dimethylformamide in the presence of lithium chloride. The product precipitated by the addition of nonsolvents to these solutions is highly amorphous and is called *regenerated cellulose*.

The rigidity of wood is due to hydrogen bonding between the cellulose molecules and lignin. When exposed to ammonia, wood begins to degrade because of the breakage of the hydrogen bonds, and the cellulose chains can be shaped. When the ammonia is washed away, new hydrogen bonds form which lock the wood into the new form. Longer exposure to a base results in a disruption of the glycosidic bonds and a permanent loss of strength.

Acids act similarly and rapidly to permanently disrupt the glycosidic bonds. Thus polysaccharides degrade to their original monosaccharide units when heated with aqueous acids. Enzymes catalyze this degradation and provide both plants and animals with a source of glucose.

14.9   Modified Cellulosics

When reacted with solutions containing nitric acid and sulfuric acid, cellulose forms various nitrated products depending on the temperature and the concentration of these reactants.

Cellulose dinitrate

Although cellulose nitrate is now used for eyeglass frames and piano key coverings, it is rapidly being replaced by less flammable materials.

Cellulose acetate was initially employed as a replacement for the more flammable cellulose nitrate. The degree of acetylation of this ester can be varied; the commercial product is 80 to 97% acetylated. Both films and fibers are made by the extrusion of solutions of cellulose acetate. The films and fibers are oriented in the direction of pull, and thus they have preferential strength in that direction.

Cellulose acetate fibers are defined by the Textile Fiber Products Identification Act as manufactured fibers in which the fiber-forming substance is cellulose acetate in which not less than 92% of the hydroxyl groups are acetylated. Cellulose triacetate is employed to make many tricot fabrics and sportswear. Cellulose triacetate textile is shrink- and wrinkle-resistant and easily washed.

Rayon is defined by the Textile Fiber Products Identification Act as a manufactured fiber composed of regenerated cellulose, as well as manufactured fibers composed of regenerated cellulose in which substituents have replaced not more than 15% of the hydrogen atoms of the hydroxyl groups.

The cuprammonia process, the viscose process, and the acetate process have been employed for the production of rayon. Cuprammonia and viscose rayons have similar chemical and physical properties. Both are easily dyed and lose their strength when wet because of a disruption of hydrogen bonding; this wet strength is improved through chemical treatment of the rayon fabrics. Acetate rayon is readily softened in the ironing process and loses its luster in boiling water.

14.10   Silicones (Siloxanes)

Silicones are organo polysiloxanes with the following repeating unit:

$$
\begin{array}{c}
R \\
| \\
+\!Si\!-\!O\!+ \\
| \\
R
\end{array}
$$

The chemical, physical, and thermal properties and resistance to degradation of polysiloxanes is the result of the high energy (106 kcal/mol) and the relatively large amount of ionic character of the siloxane bond. The ionic character of the Si—O bond facilitates acid- and base-catalyzed rearrangement and/or degradation reactions. Under inert conditions, highly purified polydiphenyl- and polydimethylsiloxanes are stable at 350 to 400 °C.

The thermal stability, as well as structure-related properties, such as resistivity and elasticity, of polysiloxanes is dependent on the nature of the pendant groups on the silicon atoms. Thus high-molecular-weight polydimethylsiloxanes are attacked at temperatures near 200 °C in the presence of oxygen, but substitution of a phenyl group for one methyl group raises the oxidative stability to 225 °C.

Extensive studies of polysiloxanes have shown that the thermal properties are also dependent on the chain length, the end groups, and the nature of the synthesis. The effect of the latter is related to the inclusion of impurities, the pore size, the amount and distribution of crystallinity, etc.

Silicones possess both thermal stability and good mechanical, chemical, and electric properties between −70 and 250 °C. In the absence of oxygen, many linear siloxanes degrade at temperatures greater than 350 °C to give cyclic products. Oxidative degradation generally occurs at lower temperatures.

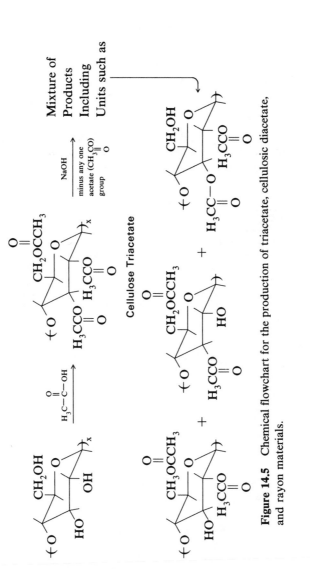

Figure 14.5 Chemical flowchart for the production of triacetate, cellulosic diacetate, and rayon materials.

The greater stability of sterically hindered siloxanes indicates that oxidation occurs at the silicon atom. Stability toward oxidative cleavage is dependent on both the nature of the organic groups and the backbone structure.

14.11   Polyphosphazenes

Polyphosphazenes are the most important and the most thermally characterized of the phosphorus-containing inorganic polymers. Linear, cyclolinear, and cross-linked cyclomatic polymers based on phosphazene structures have been produced. The repeating units of some polyphosphazenes are as follows:

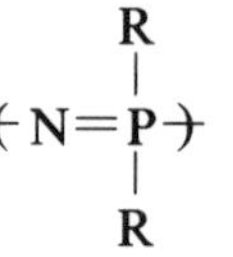

$$\begin{array}{c} R \\ | \\ \left(\!\!-N\!=\!P\!-\!\right) \\ | \\ R \end{array}$$

where R can be F, O—R, O—AR, and the following cyclolinear structure:

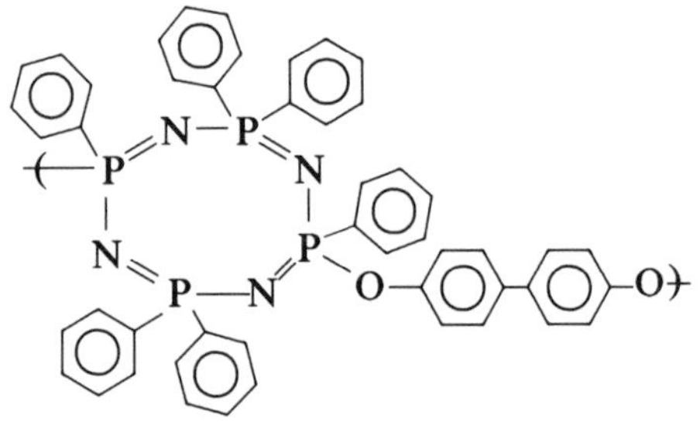

The observed low T_g's of most polyphosphazenes are consistent with the low barrier to internal rotation predicted for them and indicate the potential these polymers have for elastomeric applications. Theoretical calculations, based on rotational isomeric models assuming localized π bonding, predict the lowest ($\sim$100 cal per mol of repeating unit) known polymer barrier to rotation for the skeletal bonds of polydifluorophosphazene.

Temperature intervals between the T_g and the T_m of polyphosphazenes are unusually small and generally fall outside the frequently cited empirical relation $0.5 \leq T_g / T_m \; (K) \leq 0.67$. This behavior could be related to complications in the first-order transition generally found with organo-substituted phosphazenes. Two first-order transitions within a temperature interval of about 150 to 200 °C are usually observed for organo-substituted phosphazenes. The lower first-order transition can be detected using differential scanning calorimetry (DSC), differential thermal analysis (DTA), and thermal mechanical analysis (TMA). Examination by optical microscopy reveals that the crystalline structure is not destroyed but crystalline portions persist throughout the extended temperature range to a higher-temperature transition which appears to be the T_m, the true melting temperature. The nature of this transition behavior resembles the transformation to a mesomorphic state such as that present for nematic liquid crystals. It appears from the relation between the equilibrium melting temperature (heat of fusion H_m and entropy of fusion S_m; $T_m = H_m / S_m$) and the low value of H_m at T_m compared with the heat of fusion found for the lower temperature transition, that the upper-temperature transition T_m is characterized by a quite small entropy change compared with entropy changes found for most other polymers. It is possible that an onset of chain motion occurs between the two transitions, leading to only a small additional gain in conformational entropy at the higher-temperature transition, T_m.

The lower-temperature transition is sensitive to structural changes, and such changes usually parallel changes in the material's T_g. The T_m is generally affected much less, compared with the lower-temperature transition, by structural changes.

14.12   References

H. R. Allcock, *Heteroatom Ring Structure Polymers*, Academic Press, New York (1967).

H. R. Allcock, *Phosphorus-Nitrogen Compounds*, Academic Press, New York (1972).

C. E. Carraher, Organometallic Polymers, *J. Chem. Educ.* **58** (11), 921 (1982).

C. E. Carraher, *J. Macromol. Sci. Chem.*, **A17**(8), 1293 (1982).

H. Newroth and R. L. Hill, *The Proteins*, Academic Press, New York (1975).

E. Ott, and H. M. Spurlin, *Cellulose and Cellulose Derivatives*, Interscience, New York, (1954).

M. W. Pigman, and R. M. Gregg, *Chemistry of Carbohydrates*, Academic Press, New York 1948).

J. H. Saunders, and K. C. Frisch, *Polyurethanes*, Interscience, New York (1962).

15 | High-Performance Polymers

15.1   Introduction

Several high-performance or engineering polymers, such as the polyfluorocarbons, acetals, ABS, nylons, polyurethanes (PUs), silicones, and phosphazenes, have been described in previous chapters. Several elastomers, such as butyl rubber, EPDM (elastomeric terpolymer from ethylene, propylene, and a nonconjugated diene), and Neoprene, which play a vital role in engineering, and a host of classic thermosets should also be considered high-performance polymers. The properties of other high-performance polymers are described in this chapter.

15.2   Alkyds, Polyesters, and Allylic Resins

The term *alkyd* was originally employed to describe oil-modified polyesters, but is now used for many other polyester plastics and coatings. Glycerol can act as a difunctional reactant at moderate temperatures and thus can yield a linear polyester containing reactive hydroxyl groups (see Figure 15.1). The unreacted secondary hydroxyl groups in the linear polyester prepolymer form esters with phthalic acid at elevated temperatures. Thus the linear polyester cures when heated. The last step usually takes place after the linear resin has been applied to an object as a coating or is present in its final shape.

Many industrially important alkyd resins and plastics are polyesters which contain unsaturated groups in their intermediate, linear state.

A curable alkyd resin can be produced by the condensation of a di-

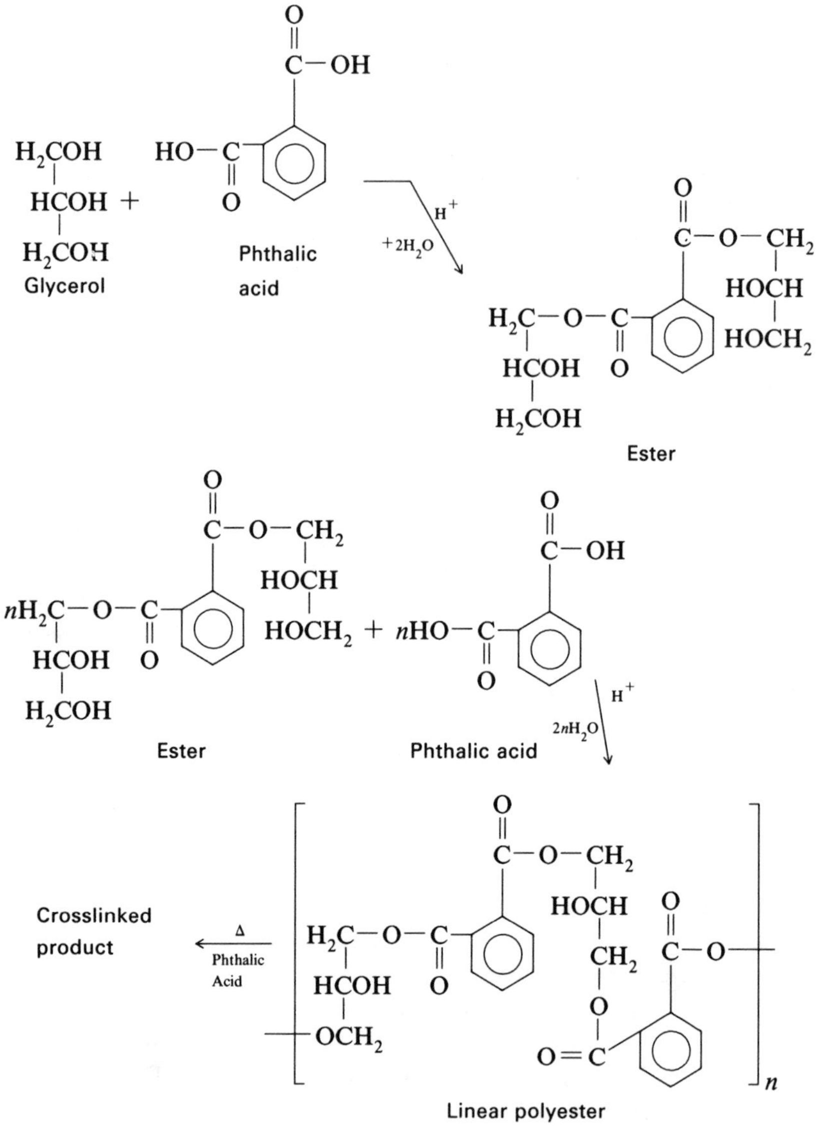

Figure 15.1.   Synthesis of cross-linked polyesters.

functional acid and a difunctional alcohol, such as phthalic acid or anhydride and ethylene glycol, in the presence of an unsaturated acid or ester. Unsaturated linear polyesters of this type, called *prepolymers*, are mixed with fillers, such as clay or ground limestone, and peroxy compounds. These premixes are sometimes used as putties, called bulk-molding compounds (BMC), or as impregnated sheets, called sheet-molding compounds (SMC).

Since it is advantageous to polymerize (cure) these mixtures at room temperature, reducing agents and initiators are usually added to accelerate the production of free radicals. The general reactions involved in the production of these polymeric compositions are shown in Figure 15.2.

Commercial polyesters are heterogeneous products consisting of some

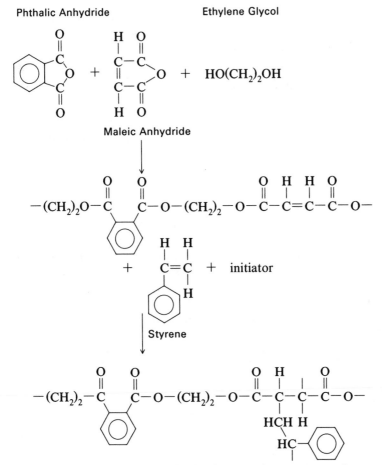

Figure 15.2. Synthesis of polyester–polystyrene networks.

polystyrene (PS) and some network polymer. Empirical polyester formulations that have been developed are acceptable for a wide variety of applications.

The compositions may be cast, laid up by hand, sprayed, or molded. The final products are used as corrugated or flat panels, helmets, electric appliances, furniture, containers, automotive housings, boats, house sidings, modular bathrooms, and chemical storage tanks.

Typical properties of fiberglass-filled alkyds are shown in Table 15.1. These products have good electric (insulative) properties and are moderately resistant to solvents and acids. Polyester fibers, based on polyethylene terephthalate (PET), are now the world's leading synthetic fibers.

Biaxially oriented PET film (Mylar) is one of the strongest films available. PET film has a tensile strength which is over 300% greater than that of films of cellophane, polyethylene, or cellulose acetate. Biaxially oriented PET is widely used as magnetic, video, and industrial tape; microfilm, wire, and cable wrapping; and packaging. It is customary to coat the packaging film with polyvinylidene chloride–co-vinyl chloride to reduce oxygen transmission. Metallized PET film is used for decorative film, labels, and decals.

Because PET has a melting point T_m of 240 °C, it is difficult to mold. The difficulty in processing and the low ductility of PET have been overcome by the production of terephthalic acid esters with higher-molecular-weight dihydric alcohols, such as butylene glycol.

These moldable thermoplastic polyesters are marketed under the trade names Celanex, Tenite, and Valox. These highly crystalline thermoplastics may be extruded or injection molded and may also be reinforced by fiberglass.

These engineering thermoplastics are used as gears, bearings, pump impellers, pulleys, pump housings, switch parts, and furniture parts.

As shown in Table 15.2, the impact resistance of polybutylene terephthalate (PBT) is greater than that of PET, but because of the increased

Table 15.1 Thermal and Physical Properties of Alkyd Plastics (Typical Values)

	Fiberglass-filled alkyds
Heat deflection temperature @ 1820 kPa (°C)	200
Maximum resistance to continuous heat (°C)	200
Coefficient of linear expansion cm/cm · °C $\times$ 10^{-5}	2.0
Compressive strength (kPa)	137,900
Flexural strength (kPa)	103,000
Impact strength (Izod: cm · N/cm of notch)	106
Tensile strength (kPa)	41,000
Elongation (%)	2
Hardness: Rockwell	E80
Specific gravity	2.1

Table 15.2 Thermal and Physical Properties of Polyaryl Esters (Typical Values)

	Polyarylate	PET	PBT
Heat deflection temperature @ 1820 kPa (°C)	175	100	65
Maximum resistance to continuous heat (°C)	150	100	60
Coefficient of linear expansion cm/cm · °C × 10^{-5}	6.5	6.5	7.0
Compressive strength (kPa)	93,000	86,000	75,000
Flexural strength (kPa)	79,000	112,000	96,000
Impact strength (Izod: cm · N/cm of notch)	215	26	53
Tensile strength (kPa)	68,000	62,000	55,000
Elongation (%)	50	100	100
Hardness: Rockwell	R125	M96	M70
Specific gravity	1.2	1.35	1.35

flexibility resulting from the additional methylene groups in the repeating units, PBT has a lower heat deflection temperature (65 °C).

The flexural modulus and heat deflection temperature of these aryl polyesters are increased by the incorporation of reinforcing fillers. PET and related aryl polyesters are resistant to nonoxidizing acids, alkalis, and salts, as well as to polar and nonpolar solvents at room temperature. (Above room temperature some alkalis and acids begin to degrade polyesters.)

Allylic plastics, which are produced by the polymerization of diallyl phthalate, have high heat deflection temperatures and high strengths. These cross-linked polyesters have solvent- and corrosion-resistant properties similar to those cited for alkyds. The properties of allylic plastics are shown in Table 15.3.

The principal synthetic polymers used as coatings are alkyd resins, styrene copolymers, polyvinyl acetate (PVAc), urea, melamine, phenolic and epoxy

Table 15.3 Thermal and Physical Properties of Allylic Plastics (Typical Values)

	Fiberglass-filled allylic plastics (DAP)
Heat deflection temperature @ 1820 kPa (°C)	200
Maximum resistance to continuous heat (°C)	150
Coefficient of linear expansion cm/cm · °C × 10^{-5}	2.0
Compressive strength (kPa)	186,000
Flexural strength (kPa)	131,000
Impact strength (Izod: cm · N/cm of notch)	106
Tensile strength (kPa)	58,000
Elongation (%)	4
Hardness: Rockwell	E80
Specific gravity	1.7

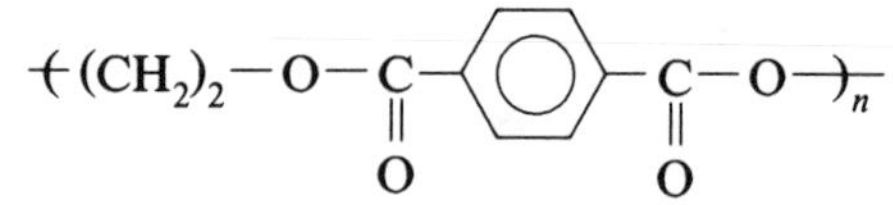

$$\{(CH_2)_2 - O - \underset{O}{\overset{\|}{C}} - \langle O \rangle - \underset{O}{\overset{\|}{C}} - O\}_n$$

Polyethylene terephthalate

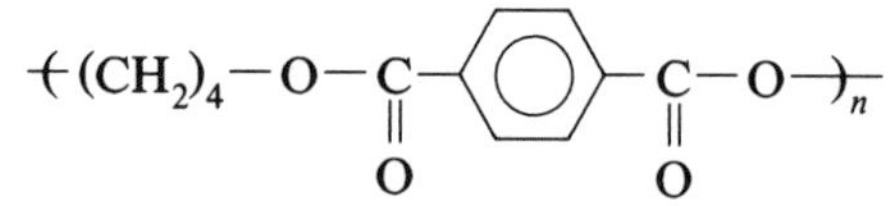

$$\{(CH_2)_4 - O - \underset{O}{\overset{\|}{C}} - \langle O \rangle - \underset{O}{\overset{\|}{C}} - O\}_n$$

Polybutylene terephthalate

Figure 15.3.   Structures of terephthalate esters.

resins, PU, and polymethyl methacrylate (PMMA). Polymer coatings consist of a resin or resin-forming compound (binder) and may contain other components, such as pigments, fillers, and solvents. However, the properties of the coating are dependent primarily on the binder. In many instances, the type of coating is indicated by the descriptive name of varnish, lacquer, latex, etc.

15.3 Phenolic Resins

Leo Baekeland obtained thermoplastic resoles (one-stage resins) by adding stoichiometric quantities of formaldehyde to phenol and heating gently under alkaline conditions. He also produced thermoplastic novolacs (two-stage resins) by using less than stoichiometric quantities of formaldehyde and heating under acid conditions.

These thermoplastic resoles and novolacs are mixed with lubricants, pigments and additives, such as wood flour. The molding compound is converted to an infusible resin by heating it under pressure in a mold. A typical sequence of chemical reactions associated with the formation of this complex, three-dimensional polymer is shown in Figure 15.4. Typical properties of phenolic plastics are shown in Table 15.4.

Cured phenolic resins have outstanding heat resistance, resistance to cold flow, good electric (insulation) properties, and good dimensional stability. Phenolic resins have good adhesive properties and are employed in the production of sandpaper, abrasive wheels, and brake linings. These resins are also used as casting resins.

Phenolic plastics are more resistant to dilute oxidizing acids than furan plastics (see Sec. 15.6), but, unlike furan plastics, phenolic plastics are swollen

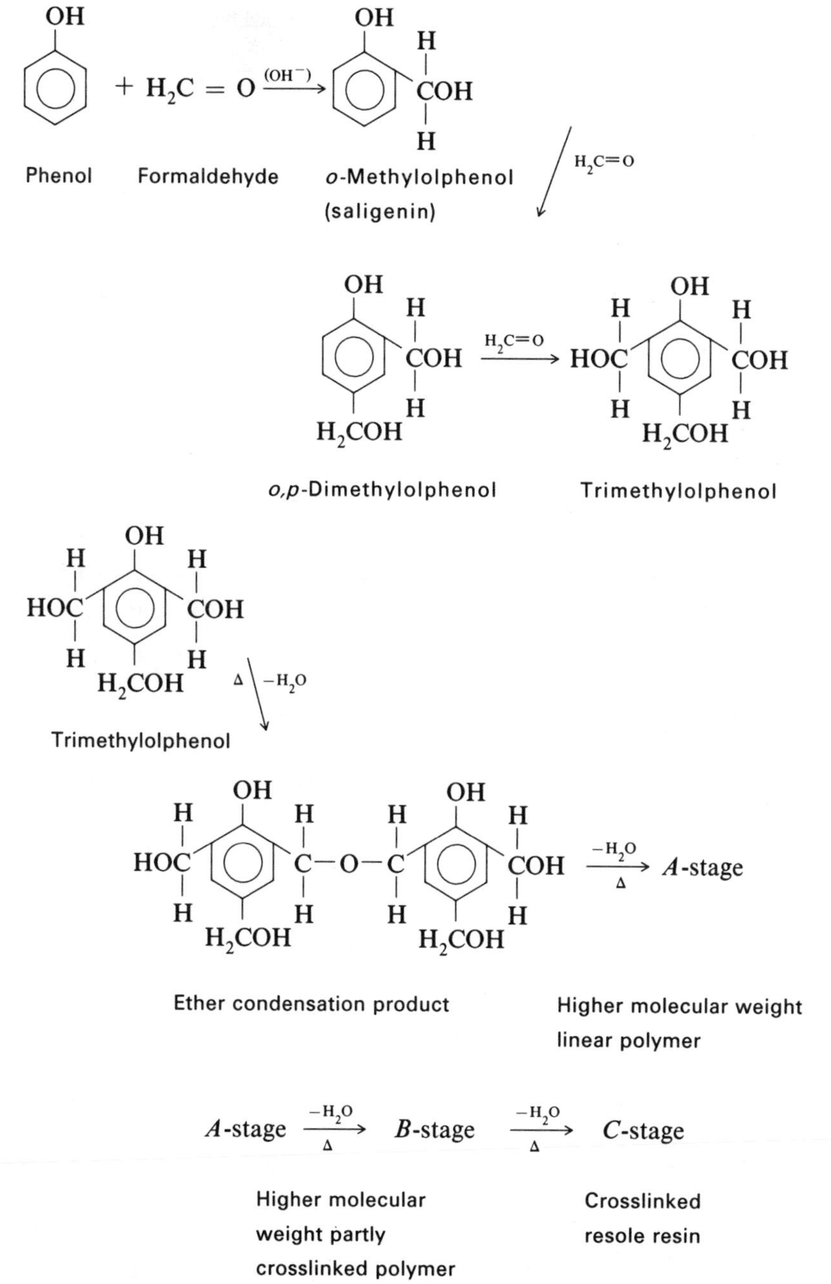

Figure 15.4 Typical sequences leading to the formation of phenol–formaldehyde resins.

Table 15.4   Thermal and Physical Properties of Phenolic Plastics
(Typical Values)

	Wood-flour-filled	Mineral-filled
Heat deflection temperature @ 1820 kPa (°C)	165	200
Maximum resistance to continuous heat (°C)	160	175
Coefficient of linear expansion cm/cm · °C × 10⁻⁵	3.0	2.0
Compressive strength (kPa)	172,000	172,000
Flexural strength (kPa)	62,000	82,000
Impact strength (Izod: cm · N/cm of notch)	21.5	21.5
Tensile strength (kPa)	48,000	41,000
Elongation (%)	0.5	0.5
Hardness: Rockwell	M100	M110
Specific gravity	1.4	1.5

by alkaline solutions. Their use in missile nose cones is based on their tendency to carbonize and produce a protective thermal barrier for the nose cones.

15.4   Amino and Amido-Derived Resins

Plastics produced by the reaction of formaldehyde and melamine or cyanuramide were introduced commercially in 1939. Today urea is also employed as a nitrogen-containing reactant. These are often discussed under the term *amino resins*. However, urea contains amide groups ($-\overset{\|}{\underset{O}{C}}NH_2$), and thus urea-formaldehyde plastics are amide-formaldehyde plastics and not amino plastics. Nevertheless, in its production and sales statistics, the U.S. Tariff Commission usually includes both urea and melamine resins under the heading of amino resins.

Many of the properties of urea plastics are similar to those of the phenolics, but, unlike phenolics, the urea plastics are not dark and are characterized by pastel and translucent colors as well as slightly superior insulating electric properties. Urea is tetrafunctional. As shown in Figure 15.5, linear and cross-linked network products are readily produced.

The principal filler used for the production of light-colored urea plastics is alpha cellulose, which has an index of refraction similar to that of the urea-

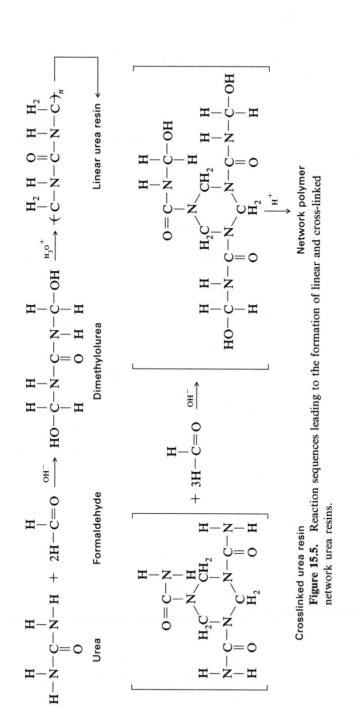

Figure 15.5. Reaction sequences leading to the formation of linear and cross-linked network urea resins.

based resins. Wood flour is used in place of alpha cellulose for the production of dark molding compounds. It is customary to mold urea plastics at 135 to 175 °C.

A typical cellulose-filled urea plastic has a tensile strength of 55,000 kPa, an Izod impact strength of 16 cm · N/per centimeter of notch, and a coefficient of linear expansion of 3×10^{-5} cm/cm · °C. Urea-formaldehyde plastics have good electric insulating properties. Unlike phenolic plastics, urea plastics do not carbonize when an electric arc is placed on their surfaces. They also have a high dielectric strength.

Melamine (MF) resins (Cymel, Melmac, Resimene) are produced by the formylation of melamine (2,4,6-triamino-1,3,5-triazine). Melamine has six active hydrogen atoms and hence forms mono-, di-, tri-, tetra-, penta-, and hexamethylol melamines. The methylol derivatives may be etherified with alcohols such as 1-butanol. A hexamethyl ether (hexamethoxymethylmelamine) is commercially available and may be used as an intermediate. Some of the reactions leading to resinification of melamine are shown in Figure 15.6.

A typical cellulose-filled melamine-formaldehyde plastic has a slightly higher tensile strength (70,000 kPa), a slightly higher coefficient of linear expansion $(4 \times 10^{-5}$ cm/cm · °C), and a higher heat deflection temperature (150 vs. 130 °C) than a typical cellulose-filled urea plastic. A melamine plastic has good nonconductance and stability with regard to electrical properties, good resistance to flame, and fair resistance to solvents and acids. Neither the melamine, the urea, nor the phenolic plastics are recommended for continuous exposure to strong alkalis.

15.5   Epoxy Resins

The terms *epoxy resin* (EP) and *ethoxyline resin* are used to describe polyphenol derivatives and other compounds which resinify by ring opening of oxirane rings. The polyphenols are typically diphenylolmethane or bisphenol A [2,2-bis(4-hydroxyphenyl)propane]; the latter diol is obtained by the condensation of phenol and acetone. Bisphenol A, is the most widely used phenol for epoxy resin production.

The most widely used epoxy resin intermediates (Araldite, Epon, and Epi-Rez) are produced from the reaction of bisphenol A and epichlorohydrin, as shown in Figure 15.7.

These resins can be cured by the addition of polyamines or by heating

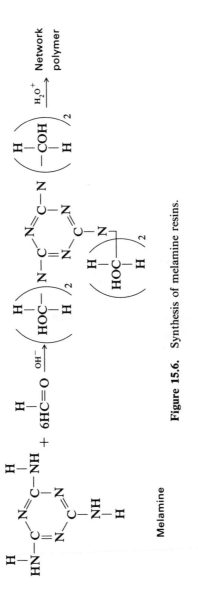

Figure 15.6. Synthesis of melamine resins.

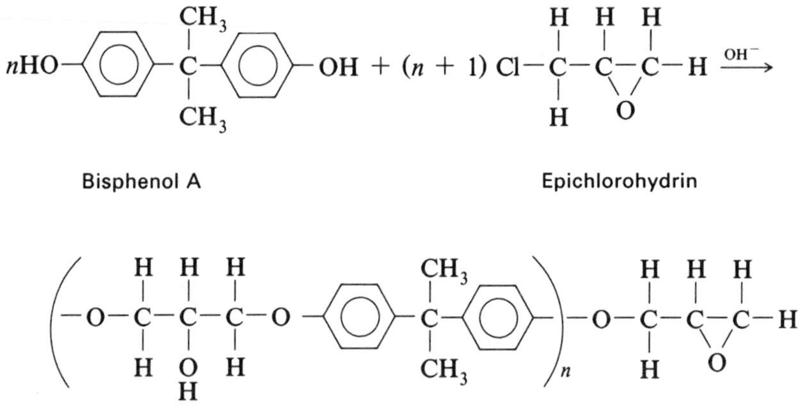

Bisphenol A   Epichlorohydrin

Figure 15.7. Production of epoxy resins.

the resins with cyclic anhydrides. Filled epoxy resins are used for the manufacture of tools for aircraft and automobiles. The liquid resins containing catalysts may be used as potting compounds, adhesives, and components of asphalt or coal tar coatings. Skidproof surfaces are produced by sprinkling coarse aggregates on partly cured epoxy resin surfaces.

As shown in Table 15.5, the epoxy plastics have fair resistance to high temperatures and have good mechanical properties. Cured epoxy resins are resistant to nonoxidizing acids, alkalis, and salts. Because of the presence of polar hydroxyl pendant groups, these polymers have good adhesion to substrates such as wood or metal.

Table 15.5 Thermal and Physical Properties of Epoxy Resins (Typical Values)

	Epoxy plastic	Fiberglass-filled	Glass-sphere-filled
Heat deflection temperature @ 1820 kPa (°C)	140	150	115
Maximum resistance to continuous heat (°C)	120	135	110
Coefficient of linear expansion cm/cm · °C × 10^{-5}	2.5	2.0	2.5
Compressive strength (kPa)	120,000	206,000	82,000
Flexural strength (kPa)	124,000	103,000	41,000
Impact strength (Izod: cm · N/cm of notch)	53	53	10
Tensile strength (kPa)	51,000	82,000	41,000
Elongation (%)	5	4	1
Hardness: Rockwell	M90	M105	—
Specific gravity	1.2	1.8	1.8

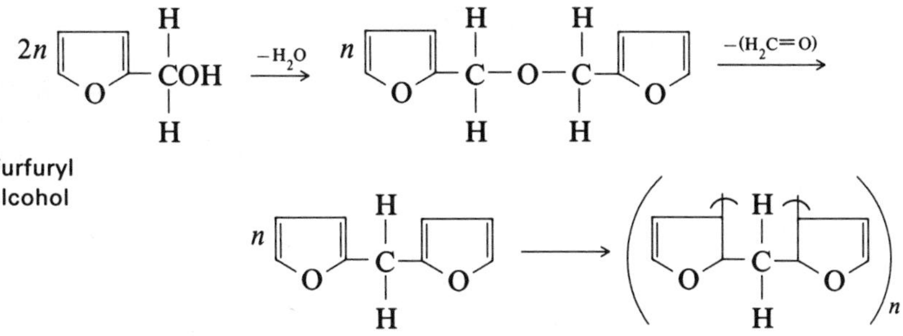

Figure 15.8. Synthesis of furan resin.

15.6 Furan Resins

Furan resins are produced by the polymerization of furfural or furfuryl alcohol in the presence of acids (see Figure 15.8). The properties of these dark-colored resins are shown in Table 15.6. Furan resins have a relatively low heat deflection temperature (80 °C) and good mechanical properties. These materials, which are widely used as jointing materials for brick and tile, are characterized by excellent resistance to nonoxidizing acids, alkalis, and salts but are affected by the presence of oxidizing acids such as nitric acid. The furan plastics are also resistant at room temperature to nonpolar solvents, such as benzene, and to polar solvents, such as ethanol.

15.7 Polycarbonates

Polycarbonates (PCs) have the following repeating unit:

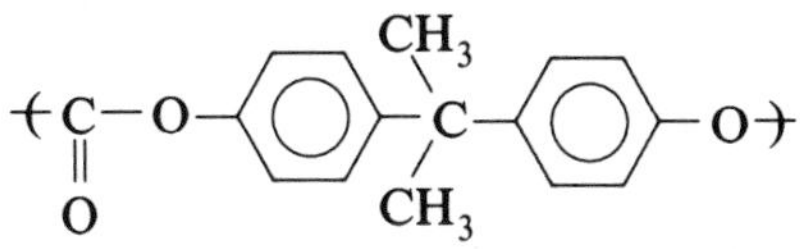

These resins are characterized by unusual toughness and clarity. As shown in Table 15.7, the addition of glass fibers improves the heat resistance of PC but reduces the impact resistance; it also reduces the clarity. PCs are hydrolyzed by water when large surfaces are available, but molded or extruded articles are resistant to water, aqueous salt solutions, and nonpolar solvents. However, because of hydrolyzable groups in the repeating units, PC is attacked by mineral acids and alkalis.

Table 15.6 Thermal and Physical Properties of Furan Plastics (Typical Values)

	Furan (coke-filled)
Heat deflection temperature @ 1820 kPa (°C)	80
Maximum resistance to continuous heat (°C)	100
Coefficient of linear expansion cm/cm · °C $\times$ 10^{-5}	7.5
Compressive strength (kPa)	69,000
Flexural strength (kPa)	34,000
Impact strength (Izod: cm · N/cm of notch)	21
Tensile strength (kPa)	41,000
Elongation (%)	1.5
Hardness: Rockwell	R110
Specific gravity	1.7

Industrial PCs are typically thermoplastics. They exhibit high impact strengths, even at low temperatures. This is attributable, in part, to a combination of relatively high order in the amorphous regions and high disorder in the crystalline regions. The crystalline and "ordered" portions of the amorphous regions provide strength. The amorphous areas and "disordered" portions of the crystalline regions contribute flexibility. PCs have good heat resistance, have good oxidative stability, and are transparent.

PCs are processed by conventional injection molding and extrusion. The impact resistance of molded or extruded PCs is decreased when they are exposed to solvents which tend to craze these plastics. Considerable care must be exercised in the extrusion process in order to obtain impact-resistant PC sheets. PCs are used for interior components of aircraft, household appliances,

Table 15.7 Thermal and Physical Properties of Polycarbonates (Typical Values)

	Unfilled	20% Fiberglass-filled
Heat deflection temperature @ 1820 kPa (°C)	130	145
Maximum resistance to continuous heat (°C)	115	130
Coefficient of linear expansion cm/cm · °C $\times$ 10^{-5}	6.8	2.2
Compressive strength (kPa)	86,000	124,000
Flexural strength (kPa)	93,000	158,000
Impact strength (Izod: cm · N/cm of notch)	534	106
Tensile strength (kPa)	72,000	131,000
Elongation (%)	110	4
Hardness: Rockwell	M70	M92
Specific gravity	1.2	1.4

lenses, switch gears, transformer housings, sports helmets, recreational vehicle bodies, propellers, body armor, and glazing.

The T_m of PCs is decreased from 225 to 195 °C when the methyl pendant groups in bisphenol A are replaced by propyl groups. This is due to the added volume occupied by the propyl groups which act against both crystallization and compact inter-chain structure.

15.8 Polyarylether Ketones

Polyarylether ketone (PEEK) has the following repeating unit:

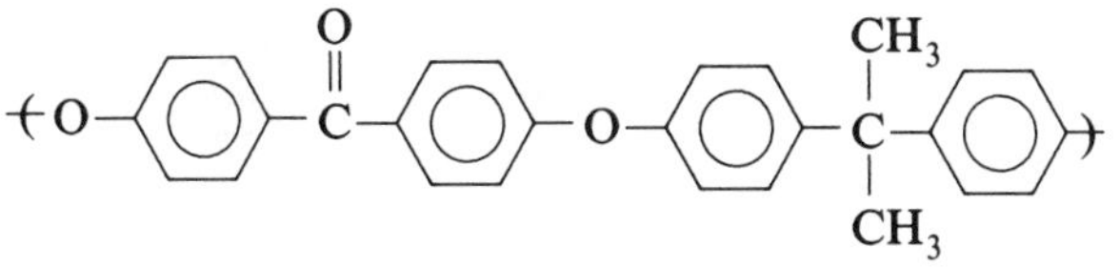

Because of the stiffening effects of the phenyl and carbonyl groups, PEEK has a high heat deflection temperature (300 °C). This high-performance polymer has high strength and excellent resistance to solvents.

15.9 Polyaryl Sulfones

Most polyaryl sulfones have the following repeating unit:

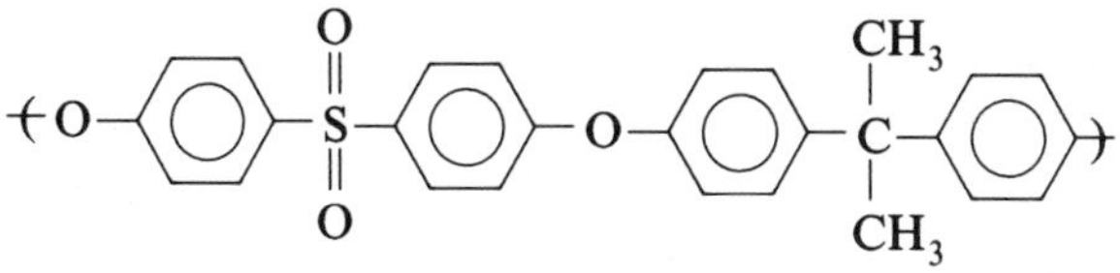

As shown in Table 15.8, polyaryl sulfones have excellent mechanical properties. The commercial products differ somewhat in properties, but all have high heat deflection temperatures (at least 175 °C). These thermoplastic engineering polymers have excellent resistance to nonoxidizing acids, salts, and alkalis, and to polar solvents. They are attacked by nonpolar solvents such as benzene.

Since the sulfone group is chain-stiffening, these amorphous plastics are rigid. However, some flexibility is provided by the ether and methylene groups. The T_g of these materials is high (> 190 °C), but polyalkyl sulfones, which

Table 15.8   Thermal and Physical Properties of Polysulfones (Typical Values)

	Polysulfone	Polyphenyl sulfone	Polyether sulfone
Heat deflection temperature @ 1820 kPa (°C)	175	205	205
Maximum resistance to continuous heat (°C)	150	165	165
Coefficient of linear expansion cm/cm · °C × 10⁻⁵	5.4	3.1	5.5
Compressive strength (kPa)	96,000	86,000	96,000
Flexural strength (kPa)	107,000	86,000	127,000
Impact strength (Izod: cm · N/cm of notch)	80	640	80
Tensile strength (kPa)	82,000	82,000	82,000
Elongation (%)	25	25	25
Hardness: Rockwell	M69	M80	M88
Specific gravity	1.24	1.29	1.37

lack the stiffening phenylene moieties, are characterized by low T_g values and poor resistance to oxidation.

Polyaryl sulfones are injection moldable or extrudable on standard processing equipment. These plastics are used for meter housings, light fixture sockets, aircraft components, household appliances, hardware, and film.

15.10   Polyphenylene Oxide and Phenoxy Resins

Polyphenylene oxide (PPO) is a highly crystalline, heat-resistant polymer with the following repeating unit:

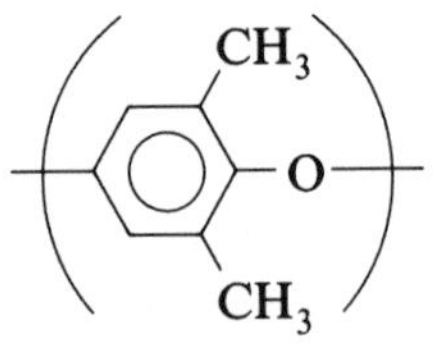

PPO has a high T_m because of the presence of the chain-stiffening 1,4-phenylene moiety, which makes up the major portion of the backbone of this polymer (the oxygen atom constitutes the remainder). Because of its very high T_m, this polymer cannot be processed by traditional extrusion or molding techniques.

PPO is formed into usable objects by long (1 to 5 min) compression molding at 350 to 425 °C. Sintering, like that observed in powdered-metal molding technology, occurs under these conditions. Because of the existence of thermal stresses in the sintered parts, it is customary to incorporate about 18% of a finely divided alumina filler in the polymer before fabrication. However, blends of PPO with PS (Noryl) are more readily moldable.

Both PPO and the PPO–PS blend have heat deflection temperatures of more than 100 °C.

PPO has excellent resistance to nonoxidizing acids, alkalis, salts, and polar solvents but is attacked by nonpolar solvents, such as benzene. As shown in Table 15.9, glass-filled PPO has a heat deflection temperature of 145 °C and excellent mechanical properties.

PPO may also be flame-sprayed and formed by high-energy forging.

Because of its inherent self-lubricating qualities, PPO is used for bearings, seals, rotors, and vanes of process pumps. It has excellent electric insulating properties.

A related material, phenoxy resin, is a thermoplastic composed of the following repeating unit:

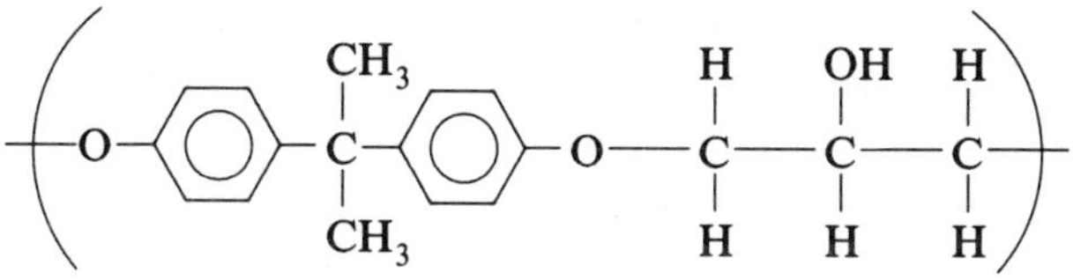

Phenoxy resins are high-molecular-weight, amorphous, linear pseudoepoxy resins that do not contain epoxy groups. Because of the presence of hydroxyl

Table 15.9 Thermal and Physical Properties of Polyphenylene Oxide (Typical Values)

	PPO	Glass-filled PPO
Heat deflection temperature @ 1820 kPa (°C)	100	145
Maximum resistance to continuous heat (°C)	80	130
Coefficient of linear expansion cm/cm · °C $\times\ 10^{-5}$	5.0	2.0
Compressive strength (kPa)	96,000	123,000
Flexural strength (kPa)	89,000	144,000
Impact strength (Izod: cm · N/cm of notch)	270	107
Tensile strength (kPa)	55,000	120,000
Elongation (%)	50	4
Hardness: Rockwell	R115	R115
Specific gravity	1.1	1.3

groups, these transparent plastics may be cured or cross-linked by diiso-cyanates or anhydrides of dicarboxylic acids.

Phenoxy resins may be extruded, injection-molded, and blow-molded to produce pipe, sporting goods, containers, and appliance housings. Since phenoxy resins are soluble in methyl ethyl ketone, they have been used as adhesives and cross-linkable protective coatings.

15.11 Polyphenylene Sulfide

Polyphenylene sulfide (PPS) is a high-melting (290 °C), injection-mold-able crystalline polymer with the following repeating unit:

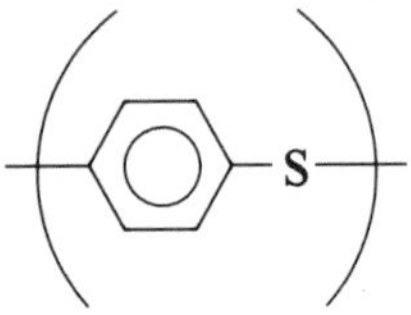

Unlike the corresponding PPO, PPS is resistant to organic solvents. PPS is sold under the trade name of Ryton. Its general properties are shown in Table 15.10.

The addition of 40% fiberglass increases the heat deflection temperature to 250 °C.

The mechanical properties of molded PPS parts are improved by heating,

Table 15.10 Thermal and Physical Properties of Polyphenylene Sulfides (Typical Values)

	Unfilled PPS	40% glass-filled PPS
Heat deflection temperature @ 1820 kPa (°C)	135	250
Maximum resistance to continuous heat (°C)	110	200
Coefficient of linear expansion cm/cm · °C $\times 10^{-5}$	5.0	2.2
Compressive strength (kPa)	110,000	144,000
Flexural strength (kPa)	96,000	207,000
Impact strength (Izod: cm · N/cm of notch)	21	75
Tensile strength (kPa)	74,000	141,000
Elongation (%)	1.1	1
Hardness: Rockwell	R123	R123
Specific gravity	1.3	1.6

which causes cross-linking of the polymer. PPS is resistant to nonoxidizing acids, alkalis, and salts, and to both polar and nonpolar solvents.

15.12   References

J. Agranoff, ed., *Modern Plastics Encyclopedia*, McGraw-Hill, New York (1980).

G. M. Bartenev, *Friction and Wear of Polymers*, Elsevier, New York (1981).

G. C. Berry and C. E. Sroog, *Rigid Chain Polymers: Synthesis and Properties*, Wiley, New York (1979).

W. Black and J. Preston, eds., *High-Modulus Wholly Aromatic Fibers*, Dekker, New York (1973).

R. A. Campbell, ed., *Advances in Polyamine Research*, Raven, New York (1978).

H. J. Cantow, *Polymer Products*, Springer-Verlag, New York (1981).

A. Ciferri and I. M. Ward, eds., *Ultra-High Modulus Polymers*, Burgess-Intl Ideas, Englewood, N.J. (1979).

R. D. Deanin, ed., *New Industrial Polymers*, American Chemical Society, Washington, D.C. (1974).

J. H. DuBois and F. W. John, *Plastics*, 5th ed., Van Nostrand-Reinhold, New York (1974).

R. T. Fenner, *Principles of Polymer Processing*, Chemical Publishing, New York (1980).

J. Frados, *SPI Plastics Engineering Handbook*, 4th ed., Van Nostrand-Reinhold Co., New York (1976).

C. A. Harper, ed., *Handbook of Plastics and Elastomers*, McGraw-Hill, New York (1975).

R. V. Milby, *Plastics Technology*, McGraw-Hill, New York (1973).

D. C. Miles and J. H. Briston, *Polymer Technology*, Chemical Publishing, New York (1979).

R. B. Seymour, *Modern Plastics Technology*, Reston Publishing, Reston, Va. (1975).

G. A. Stahl and D. S. Garner, *Polymers for Hostile Environments*, ACS Symposium Series, Washington, D.C. (1983).

J. L. White, ed., *Fiber Structure and Properties*, Interscience, New York (1979).

16 | Selection of Polymers for Special Applications

16.1 Introduction

Because of their different structures, polymers may be used for a wide variety of applications, ranging from enhanced oil recovery to components for aerospace vehicles. Fortunately, one can predict the properties of polymers, to some extent, from a knowledge of their structure.

A brief survey of some current growth areas for special polymer applications is provided in this chapter. Additional information may be found in the references cited at the end of this chapter.

16.2 Water-Soluble Polymers

The rules governing the water solubility of polymers are similar to those governing the water solubility of smaller organic molecules except that the extent of polymer solubility and the range of polymeric structures are more limited. Selected commercially available water-soluble polymers are shown in Figure 16.1.

The presence of highly electronegative atoms which can participate in hydrogen bonding is required for the solubility of polymers in water. Such groups include amines, imines, ethers, alcohols, sulfates, carboxylic acids and associated salts, and, to a lesser extent, thiols. The water solubility is also affected by pH and the formation of charged species. Thus the copolymer derived from vinylamine and vinyl sulfonate is not soluble in water, whereas the corresponding sodium salt of this copolymer is water-soluble.

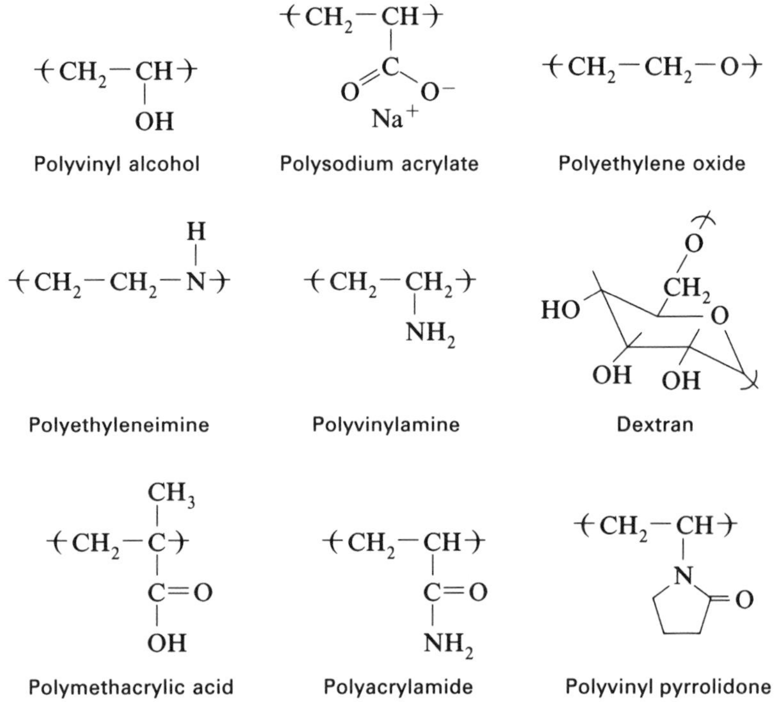

Figure 16.1. Structures of water-soluble polymers.

The amount and the rate of water solubility or swelling can be decreased by cross-linking and the substitution of nonpolar units for polar units in the polymer. Most water-soluble polymers possess both hydrophobic and hydrophilic moieties. This combination of hydrophobic and hydrophilic moieties affects the shape of the polymer chains in solution. Thus many water-soluble polymers in aqueous solution exist as random or partially helical chains which are partially extended to allow fuller hydrogen bonding with the water molecules.

Since the polar groups repel each other, the expanded random coil molecules tend to become stiff rods. The nonpolar portions of the water-solubilized polymer face toward the organic phase at the organic–aqueous interfaces, and the polar portions preferentially point away from the organic phase.

When a sufficient number of polar groups are present, linear polymers may be soluble in water. Of course, this solubility may be hindered if the polar groups on one chain are attracted strongly to polar groups on an adjacent chain. Thus in spite of a preponderance of hydroxyl groups, cellulose is swollen by water but does not dissolve because of a small degree of cross-linking and the presence of interchain and intrachain hydrogen bonding.

The water absorption properties of cellulose may be increased by grafting on polar groups, such as carboxyl groups, as in "Super Slurpers," and this absorption may be decreased by the introduction of nonpolar moieties, such as ester groups, as in cellulose acetate.

Cellulose may be converted from a water-insoluble polymer to a water-soluble polymer by the partial etherification of some of the hydroxyl groups by dimethyl sulfate. When the degree of substitution (DS) is 1.5 to 2.0, the hydrogen bonds are sufficiently weakened, and the methylcellulose is soluble in water. Carboxymethyl ethers, such as carboxymethylcellulose (CMC), are also water-soluble. The degree of solubility is related to the DS of the polymer and the pH of the solvent.

Because of the presence of ether groups and terminal hydroxyl groups, polyethylene oxide (PEO) is also soluble in water. Likewise, polyvinyl alcohol (PVA), which contains a hydroxyl group on most of the alternate carbon atoms in the linear chain, is soluble in water.

Polyvinyl methyl ether and polymers of acrylic and methacrylic acid are also water-soluble. Because of salt formation in the polymers, the acrylic and methacrylic acid polymers, as well as CMC, are more soluble in alkaline solutions than they are in water.

High-molecular-weight linear water-soluble polymers enhance the flow of water. Thus they are added in small amounts to water used for extinguishing fires and for enhanced oil recovery.

Polyhydroxyethyl methacrylate (PHEMA) is more soluble in water than polymethyl methacrylate (PMMA). Cross-linked PHEMA, which swells but does not dissolve in water, is used in soft contact lenses.

16.3   Oil-Soluble Polymers

Like dissolves like, and this is true with both polymers and smaller molecules. Thus linear amorphous polymers with nonpolar groups are typically soluble in nonpolar solvents with solubility parameter values within 1.8 H of that of the polymer. Thus polyisobutylene (PIB) is soluble in hot lubricating oils, and small amounts of high-molecular-weight PIB are used as viscosity improvers.

If PIB or other related polymers are not soluble in the oil at room temperature, they have little effect on the viscosity of the oil under these conditions. However, if these polymers are soluble in hot oil, they will typically increase the viscosity of the hot oil and can be used as viscosity improvers.

Polystyrene (PS) is too insoluble to serve as a viscosity improver for lubricating oil, but polycyclohexylstyrene is sufficiently soluble to be used for this application. Likewise, PMMA is too insoluble to serve as a viscosity improver, but polylauryl methacrylate serves as an excellent viscosity improver in lubricating oils.

16.4   Oil-Insoluble Polymers

Polymers with solubility parameters differing from those of the solvent by at least 2.0 H, will not dissolve in the solvent at room temperature. Thus although unvulcanized natural rubber (NR), unvulcanized styrene–butadiene elastomer (SBR), unvulcanized butyl rubber, and EPDM dissolve in gasoline or benzene, the vulcanized (cross-linked) polymers are swollen but will not dissolve due to the presence of the crosslinks.

The incorporation of polar groups in unvulcanized polymers reduces their solubility in benzene. Thus the copolymer of acrylonitrile and butadiene (NBR), polychlorobutadiene (Neoprene), and fluorinated EP (the copolymer of ethylene and propylene) are less soluble in benzene and lubricating oils than the previously cited elastomers. Likewise, silicones and phosphazene elastomers, as well as elastomeric polyfluorocarbons, are insoluble in many oils and aromatic hydrocarbons because of their extremely low solubility parameters (silicons: 7–8 H; polytetrafluoroethylene: 6.2; benzene: 9.2; toluene: 8.9; pine oil: 8.6).

16.5   Flame-Retardant Polymers

Polymeric hydrocarbons, such as polyolefins, are readily combustible and can actually serve as a fuel source. In contrast, polytetrafluoroethylene (PTFE) does not burn in air but burns in oxygen or in nitrogen–oxygen mixtures which have a very high oxygen concentration.

The combustion tendency of polymers in air may be reduced by the incorporation of flame retardants, such as alumina trihydrate (ATH), which releases steam when heated, or chlorinated organic compounds and antimony oxide, which produce antimony chlorides when heated together.

Another rule of thumb concerns the oxidation tendency of polymers. Carbonates and phosphates with the CO_2 and OPO_3 moieties are already largely oxidized. Thus portions of polymers that contain the carbonate and phosphate molecules do not contribute to combustion. In fact, polyphosphate and polyphosphonate esters

$$\begin{matrix} & & O & & \\ & & \parallel & & \\ +O & - & P & - R - O + \\ & & | & & \\ & & O & & \\ & & R' & & \end{matrix}$$

are utilized as flame-retardant additives for certain clothing and textiles. The presence within polymer chains of halides also assists the material to resist ready oxidation and helps to retard combustion.

16.6 Flexible Polymers

Polymers having many flexibilizing groups (CH_2, O), such as polyethylene, polyisoprene, and polysiloxanes (silicones), are flexible. Other less-flexible polymers may be flexibilized by the introduction of flexibilizing groups. For example, polybutylene terephthalate (PBT) is more flexible than polyethylene terephthalate (PET), and nylon 11 is less rigid than nylon 6.

Flexible pendant groups also promote flexibility in polymers. Thus polyoctyl methacrylate is much more flexible than PMMA. Stiffness resulting from crystallinity may be overcome by copolymerization which reduces the tendency to crystallize. Thus although HDPE is relatively stiff, EP is flexible.

Intractable polymers, such as polyvinyl chloride (PVC), may be flexibilized, to some extent, by the formation of copolymers, such as the copolymers of vinyl chloride and vinyl acetate or octyl acrylate, or by the addition of nonvolatile low-molecular-weight compounds (plasticizers) having solubility parameters similar to those of the polymer. Thus PVC is plasticized by the addition of dioctyl phthalate. The flexibility of these products is proportional to the amount of plasticizer added. Copolymers, such as the vinyl chloride–vinyl acetate copolymer, require less plasticizer to obtain the same degree of flexibility.

16.7 Water-Repellent Polymers

In contrast to water-soluble polymers, such as polyacrylamide, which has a relatively high critical surface tension (35 dyne/cm), water-repellent polymers, such as the silicones and PTFE, have relatively low critical surface tensions (24 and 19 dyne/cm, respectively). The presence of hydroxyl groups in polymers, such as polyvinyl alcohol and polyacrylic acid which tend to hydrogen bond, also leads to high critical surface tension values for such polymers.

Other polar groups also contribute toward relatively high critical surface tensions through formation of polar-polar secondary interactions. Thus although HDPE has a critical surface tension of 31 dyne/cm, PVC has a critical surface tension of 37 dyne/cm. In contrast, the critical surface tension of polyvinyl fluoride (PVF) is 28 dyne/cm.

16.8 Heat-Resistant Polymers

The presence of stiffening groups, such as phenylene groups, in the polymer chain tends towards a decrease in chain flexibility and an increase in the glass transition temperature, T_g, of the polymer. Other factors, such as the presence of strong intermolecular forces, such as hydrogen bonding and the lack of flexible pendant groups, also contribute to the heat resistance of polymers.

Thus because of the presence of the amide stiffening group and hydrogen bonding, nylon 66 has a T_g of about 60 °C. In contrast, the T_g of hexamethylene adipate is about −70 °C.

The presence of a phenylene group in these polyamides and polyesters increases the heat resistance dramatically. Thus the T_g of polyethylene terephthalamide (aramid) is 300 °C and that of PET is 265 °C. Poly-p-oxybenzoate (Ekonol) does not melt but decomposes at 480 °C.

High-performance polymers, such as polyphenylene sulfide (PPS) (Ryton), have excellent resistance to elevated temperatures, and this property can be enhanced by the incorporation of reinforcing fibers, such as graphite fibers.

16.9 Polymers with Resistance to Gaseous Permeation

The rate of transmission of gases and vapors through polymeric films varies with the structure of both the diffusate molecule and the polymer. Polymers with polar groups, such as cellulose and cellulose acetate, are permeable to water vapor, but polymeric hydrocarbons, such as PIB, are essentially impervious to water vapor.

In contrast, cellulose and PVA are essentially impervious to gaseous alkanes, such as propane, but polyethylene and PS are pervious to these gases.

The permeability coefficient P for relatively inert gases, such as oxygen and nitrogen, is dependent on a number of factors including polymer crystallinity (where permeability decreases with an increase in polymer crystal-linity; thus permeability is moderately high for LDPE, lower for HDPE, and extremely low for polyvinylidene chloride (PVDC) and polarity of the polymer (with permeability of largely nonpolar gases decreasing as the polarity of the polymer increases; thus butyl rubber is resistant to the permeation of oxygen and nitrogen, and this resistance is greater for chlorobutyl and bromobutyl rubbers).

16.10 Transparent Polymers

Although crystalline polymers, such as PTFE, are opaque, some amorphous polymers, such as PMMA, polysulfones, and PCs, have a high degree of transparency.

Most commercial acrylic sheets have an index of refraction of about 1.5 and a total luminous transmittance of about 90%. PC sheet has an index of refraction of about 1.6 and a total luminous transmittance of about 88%. Polysulfone sheet (Udel) has an index of refraction of about 1.63 and a total luminous transparency of 80%.

16.11 Insulating Polymers

Polymers such as polyethylene, which do not have polar groups, are excellent insulators of heat and electricity. The thermal insulating properties may be improved by foaming or by the incorporation of hollow glass spheres (syntactic foams). A low-density polyethylene foam will have a thermal conductivity in the order of 0.3 BTU/ft² · h · °F · in.

The electric and heat conductivity of polymers may be increased by the incorporation of conductive fillers, such as aluminum flakes or metallic fibers.

16.12 Adhesive Polymers

Adhesives can be considered to be coatings between two surfaces. Adhesion may be defined as the process that occurs when a solid and a movable material (usually in liquid form) are brought together to form an interface, and the surface energies of the two substances are transformed into the energy of the interface.

A unified science of adhesion is still being developed. Adhesion can result from mechanical bonding between the adhesive and adherend and/or primary and/or secondary chemical forces. Contributions from chemical forces are often more important and illustrate why nonpolar polymeric materials, such as polyethylene, are difficult to bond, whereas polar materials, such as polycyanoacrylates, are excellent adhesives. There are numerous types of adhesives including solvent-based, latex, pressure-sensitive, reactive, and hot-melt adhesives.

At least one of two factors must be in operation for good adhesion. The first factor is the formation of primary and secondary bonds between the adhesive and the adherend surfaces. The primary bonds may be formed by the addition of cross-linking agents. The formation of secondary bonds sufficient to firmly bond two surfaces together usually requires groups that can form hydrogen and polar bonds. Thus polyethylene and PTFE are difficult to bond since they are nonpolar, whereas the active groups in many adhesives are polar, as in the case of the cyanoacrylic-based glues. A second factor is the formation of mechanical bonding through penetration by the bonding agent into the surface pores with subsequent hardening of the bonding agent. Polymer formation can occur prior to or subsequent to the application of the adhesive.

Adhesives differ in mode of delivery and type of material employed. Adhesives may be classified according to the following application techniques.

Solvent-Based Adhesives—In these the adhesive flows because it is dissolved in an appropriate solvent, and solidification occurs on evaporation of the solvent. Good bonds are usually formed if the solvent attacks or actually dissolves some of the plastic adherend to produce a solvent-welded bond.

Latex Adhesives—These adhesives are based on polymer lattices in which the polymers are near their T_g values so that they can flow to provide good surface contact on evaporation of the water. It is not surprising that the same polymers that are useful as latex paints are also useful as latex adhesives. Latex adhesives are widely employed for bonding pile to backings of carpets.

Pressure-Sensitive Adhesives—These are really viscous polymers which melt at room temperature, so the polymers must be used at temperatures above their T_g values to permit rapid flow. The adhesives flow because of the application of pressure. When the pressure is removed, the viscosity of the polymer is high enough to retain its adhesion to the surface. Many adhesive tapes are of this type; the back is smooth and coated with a nonpolar coating which does not bond with the "sticky" surface.

Hot-Melt Adhesives—Thermoplastics often form good adhesives simply by melting, followed by subsequent cooling after the plastic has filled surface voids. Electric "glue guns" typically operate on this principle.

Reactive Adhesives—These adhesives are either low-molecular-weight polymers or monomers which solidify by polymerization and/or cross-linking reactions after application. Cyanoacrylates, phenolics, silicones, and epoxies are examples of this type of adhesive. Plywood is formed by the impregnation of thin sheets of wood with resin. The impregnation occurs after the resin is placed between the sheets of wood.

The type of adhesives can also be classified according to the materials employed.

Phenolic resins produced by the reaction of phenol and formaldehyde were used as adhesives by Leo Baekeland in the early 1900's. This inexpensive resin is still used for binding thin sheets of wood to produce plywood. Urea resins produced by the reactions of urea and formaldehyde have been used since 1930 as binders for wood chips in particleboard.

Unsaturated polyester resins have replaced lead for auto body repair, and polyurethanes (PUs) are being used to bond polyester cord to rubber in tires, to bond vinyl film to particleboard, and as industrial sealants. Epoxy resins are used in automotive and aircraft construction and as a component of plastic cements.

Otto Bayer produced PUs in the 1940's by reacting diisocyanates, such as tolyl diisocyanate, with dihydric alcohols, such as ethylene glycol. In another experiment he added a diisocyanate to cure synthetic rubber (SR) containing hydroxyl groups. The rubber was cured (vulcanized), but it stuck to the mold. Variations of products from these two experiments are now used as two-component adhesives for bonding footwear and automotive plastic parts.

Solutions of NR have been used for laminating textiles for over a century. The mackintosh raincoat consists of two sheets of cotton adhered by an inner layer of natural rubber.

Pressure-sensitive tape (Scotch tape) consisting of a coating of a solution of a blend of NR and an ester of glycerol and abietic acid (rosin) on cellophane was developed over a half century ago. More recently, NR latex and SR have been used in place of the NR solution.

SBR is now used as an adhesive in carpet backing and packaging. Neoprene (polychloroprene) may be blended with a terpene or a phenolic resin and used as a contact adhesive for shoes and furniture. Contact adhesives are usually applied to both surfaces, which are then pressed together at a later time.

Liquid copolymers of butadiene and acrylonitrile with carboxyl end groups are used as contact adhesives in the automotive industry. Polysulfide elastomers, produced by the reaction of ethylene dichloride and sodium polysulfide; butyl rubber; a copolymer of isobutylene; and silicones, such as polydimethylsiloxane,

$$\begin{array}{c} CH_3 \\ | \\ \text{(-}O-Si\text{-)} \\ | \\ CH_3 \end{array}$$

are used as sealants and caulking cements.

One of the most interesting and strongly bonded groups of adhesives are polymers based on cyanoacrylate (Super Glue, Krazy Glue). These monomers, such as butyl-α-cyanoacrylate, polymerize spontaneously in the presence of moist air to produce excellent adhesives. These adhesives have both cyano and ester polar groups in the repeating units of the polymer chain and are used in surgery and for mechanical assemblies.

16.13 Coatings

The traditional concepts of the fundamental purposes of coatings—being decorative and protective—are being replaced by futuristic concepts of coatings as energy-collective devices, burglar alarm systems, etc. Even so, the problems of coatings' adhesion, weatherability, permeability, corrosion inhibition, flexural strength, service life, application preparation and procedures, etc., continue to be major issues in polymer science. Effective coatings generally yield good, tough, flexible films with moderate-to-good adhesion to the surface.

16.14 Corrosion-Resistant Polymers

In general, polymers are much more resistant to corrosives than metals. Most water-insoluble polymers are not attacked by aqueous salt solutions, and polymers without hydrolyzable or reactive groups are resistant to most nonoxidizing acids and alkalis.

PTFE is resistant to solvents, oxidizing acids, nonoxidizing acids, alkalis, and aqueous salt solutions. Other polyfluorocarbons, such as polyvinylidene fluoride (PVDF), are slightly less resistant to corrosives but are much more resistant than most other construction materials.

Since PVC, HDPE, and polyesters are resistant to most nonoxidizing corrosives and are less expensive and more readily processed than the polyfluorocarbons, they may be used in many instances as corrosion-resistant polymers.

16.15   References

N. M. Bikales, ed., *Water Soluble Polymers*, Plenum Press, New York (1973).

A. Blumstein, ed., *Liquid Crystalline Order in Polymers*, Academic Press, New York (1978).

A. R. Blythe, *Electrical Properties of Polymers*, Cambridge Univeristy Press, Cambridge (1980).

J. W. Boretos, *Concise Guide to Biomedical Polymers: Their Design, Fabrication and Molding*, C. C. Thomas, Springfield, Ill. (1973).

H. J. Cantow, *Specialty Polymers*, Springer-Verlag, New York (1981).

C. Carraher, J. Sheats, and C. Pittman, eds., *Organometallic Polymers*, Academic Press, New York (1978).

C. Carraher and L. Sperling, eds., *Polymer Applications of Renewable Resource Materials*, Plenum Press, New York (1983).

A. Ciferri and W. R. Krigbaum, eds., *Polymer Liquid Crystals*, Academic Press, New York (1982).

R. D. Deanin, *New Industrial Polymers*, American Chemical Society, Washington, D.C. (1974).

K. Dusek, ed., *Polymer Networks*, Springer-Verlag, New York (1982).

H. G. Elias, ed., *New Commercial Polymers*, Gordon and Breach, New York (1977).

C. Gebelein and F. Koblitz, eds., *Biomedical and Dental Applications of Polymers*, Plenum Press, New York (1981).

E. P. Goldberg and A. Nakajima, *Biomedical Polymers: Polymeric Materials and Pharmaceuticals for Biomedical Use*, Academic Press, New York 1980).

M. A. Golub and J. A. Parker, eds., *Polymeric Materials for Unusual Service Conditions*, Krieger, Melbourne, Fla. (1973).

A. Hebeisch and J. T. Guthrie, *The Chemistry and Technology of Cellulose Copolymers*, Springer-Verlag, New York (1981).

H. H. Jellinek, *Aspects of Degradation and Stabilization of Polymers*, Elsevier, New York (1978).

D. Klempner and K. Frisch, eds., *Polymer Alloys II: Blends, Blocks, Grafts and Interpenetrating Networks*, Plenum Press, New York (1980).

R. J. Kostelnik, *Polymeric Delivery Systems*, Gordon and Breach, New York (1978).

E. L. Kukacka, ed., *Applications of Polymer Concrete*, ACI, Detroit, Mich. (1981).

L-H. Lee, ed., *Advances in Polymer Friction and Wear*, (five volumes), Plenum Press, New York (1974).

M. Lewin and S. M. Atlas, *Flame-Retardant Polymeric Materials*, Vol. 3, Plenum Press, New York (1982).

J. A. Manson and L. H. Sperling, *Polymer Blends and Composites*, Plenum Press, New York (1976).

D. Meier, ed., *Block Copolymers*, Harwood, New York (1981).

Y. L. Meltzer, *Water Soluble Polymers*, Noyes, Park Ridge, N.J. (1981).

J. Mort and G. Pfister, *Electrical Properties of Polymers*, Wiley, New York (1982).

I. Perepechko, *Low-Temperature Properties of Polymers*, Pergamon Press, Elmsford, N.Y. (1981).

J. S. Robinson, ed., *Fiber-Forming Polymers: Recent Advances*, Noyes, Park Ridge, N.J. (1980).

S. L. Rosen, *Fundamental Principles of Polymeric Materials*, Wiley, New York (1982).

D. Seanor, ed., *Electrical Properties of Polymers,* Academic Press, New York (1983).

R. B. Seymour, *Hot Organic Coatings,* Reinhold, New York (1960).

R. B. Seymour, *Plastics vs. Corrosives,* Wiley, New York (1982).

R. B. Seymour, *Conductive Polymers,* Plenum Press, New York (1981).

R. B. Seymour and C. E. Carraher, *Polymer Chemistry: An Introduction,* Dekker, New York (1981).

D. O. Shah and R. S. Schechter, eds., *Improved Oil Recovery by Surfactant and Polymer Flooding,* Academic Press, New York (1977).

L. H. Sperling, *Interpenetrating Polymer Networks and Related Materials,* Plenum Press, New York (1981).

M. Szycher and W. J. Robinson, eds., *Synthetic Biomedical Polymers: Concepts and Applications,* Technomic, Westport, Conn. (1980).

H. Ulrich, *Introduction to Industrial Polymers,* Macmillan, New York (1982).

E. J. Vandenberg, ed., *Polyethers,* American Chemical Society, Washington, D.C. (1975).

* | Appendix: Acronyms and Abbreviations

A	Atomic contributions to polarization; Area; Wavelength of light
ABS	Copolymer of acrylonitrile, butadiene, and styrene
ac	Alternating current
a	Shift Factor; Area
a_c	Coefficient of expansion
ASTM	American Society for Testing and Materials
at	Atactic
ATH	Alumina trihydrate
B	Boron filaments
C	Partial volume; Capacity; Concentration; Compliance; Stress-optical coefficient
CPVC	Chlorinated polyvinyl chloride
D	Diffusion, diffusion coefficients; Density
d	Diameter; Density
dc	Direct current
DOP	Dioctyl phthalate
DP	Degree of polymerization
DS	Dielectric strength; Degree of substitution
DSC	Differential scanning calorimetry
DTA, (DT)	Differential Thermal Analysis
E	Young's modulus; Electronic contributions to polarization
E_a	Activation energy
EMI	Electromagnetic interference
EP	Polyethylene-co-propylene
EPDM	Polyethylene-co-propylene (cross-linked with a diene)
F	Fillers, reinforcements; Weight of diffusate crossing a unit area per unit time; Force
f	Distribution of fibers; Aspect ratio; Filler; Frequency of oscillation; Coefficient of expansion of filler
G	Shear modulus; Gibbs free energy
H	Heat of reaction, change
H	Hildebrand unit, solubility parameter value
HDPE	High-density (linear) polyethylene
HDT	Heat deflection temperature

HIP	High-impact polystyrene
I	Interfacial contribution to polarization
it	Isotactic
K	Kelvin
K	Bulk modulus
k	Polarizability; Absorption index
l, L	Length
L, L_0	Gauge length
LDPE	Low-density (branched) polyethylene
LLDPE	Linear LDPE
ln	Natural logarithm
LOI	Limiting oxygen index
M	Molecular weight; Chain stiffness factor
M	Matrix
M	Modulus, stiffness
n	Refractive index
N	Power loss; Number of cycles; Number of alternating loads causing failure
NBR	Polybutadiene-co-acrylonitrile
NR	Natural rubber
P	Polarizability; Permeability; Poisson's ratio; Dielectric polarization
p	Phase angle between stress and strain; Specific resistance
PA	Polyamide (nylon)
PAN	Polyacrylonitrile
PBD	Polybutadiene
PBI	Polybenzimidazole
PC	Polycarbonate
PCTFE	Polychlorotrifluoroethylene
PE	Polyethylene
PEEK	Polyarylether ketone
PEO	Polyethylene oxide
PET	Polyethylene terephthalate
PF	Phenol–formaldehyde resin
PHEMA	Polyhydroxyethyl methacrylate
PI	Polyimide
PIB	Polyisobutylene
PMA	Polymethyl acrylate
PMMA	Polymethyl methacrylate
P_0	Dipole contribution to polarization
POM	Polyacetals; Polyoxymethylene
PP	Polypropylene
PPO	Polyphenylene oxide
PPS	Polyphenylene sulfide
PS	Polystyrene
PTFE	Polytetrafluoroethylene; Teflon
PU	Polyurethane
PVA	Polyvinyl alcohol
PVAc	Polyvinyl acetate
PVB	Polyvinyl butyral
PVC	Polyvinyl chloride
PVDC	Polyvinylidene chloride
PVDF	Polyvinylidene fluoride

PVF	Polyvinyl formal; Polyvinyl fluoride
PVP	Polyvinyl pyrrolidone
Q	Charge (electric); Heat flow
R	Universal gas constant; Stress relaxation modulus; Weight change per day; Resistance
R_0	Refraction
S, s	Stress; Solubility; Solubility coefficient; Scattering coefficient; Entropy
S_0	Amplitude of stress response
SAN	Polystyrene-co-acrylonitrile
SBR	Copolymer of butadiene and styrene
SMA	Copolymer of styrene and maleic anhydride
st	Syndiotactic
T	Temperature; Direct transmission factor; Thickness
t	Time; Thickness
T_b	Boiling point
T_g	Glass transition temperature
T_m	Melting transition temperature
TBA, (TB)	Torsional braid analysis
TD	Thermal diffusion
TGA, (TG)	Thermal gravimetric analysis
TMA, (TM)	Thermal mechanical analysis
TPX	Poly-4-methylpentene-1
UF	Urea–formaldehyde resin
V	Voltage; Fractional volume; Volume
w	Weight; Width
α	Proportional; Shift factor
γ	Poisson's ratio; Surface tension; Strain
λ	Thermal conductivity; Wavelength of light
η	Viscosity, coefficient of viscosity; Index of refraction
ϵ	Dielectric constant; Strain
ρ	Density
ϕ	Partial volume
δ	Solubility parameter

Acronyms and abbreviations are not used to confuse the student nor to save printing space, but rather because it is often industrial practice to employ these abbreviations. Thus it is to the reader's advantage to do likewise. A fuller listing appears in *Polymer News* **9** (4), 101-110 (1983).

* | Index

Note a. italicized page numbers note appearance of a definition of that term
   b. boldfaced page numbers note appearance of a structure of that compound